European Space Policy

Space policy is at the cutting edge of current EU policy development and is a fascinating object of study, involving multiple and diverse actors. It is also an original and contemporary lens for studying European policy-making.

This book explores advances in European space policy and their significance for European integration. Using a 'framing' methodology, it addresses central questions in European studies to form an interdisciplinary bridge between current research in space policy and contemporary European political studies. It assesses the interests of EU institutions in space and how these institutions perceive space policy. Furthermore, it demonstrates that space is a cross-cutting policy domain affecting a diverse range of EU policy fields, such as security, transport and migration, and underpinning the twenty-first century European and global economy. In doing so, this book firmly locates space policy in the field of European studies.

This innovative book will be of key interest to students and scholars of a range of policy areas, including common foreign and security policy, technology policy, transport policy, internal market policies, development aid and disaster-risk management.

Thomas Hörber is Full Professor at the EU–Asia Institute, ESSCA – School of Management, Angers, France.

Paul Stephenson is Assistant Professor at the Department of Political Science, Faculty of Arts and Social Sciences, Maastricht University, the Netherlands.

Routledge advances in European politics

European Space Policy

European integration and the final frontier

**Edited by Thomas Hörber
and Paul Stephenson**

Routledge
Taylor & Francis Group

LONDON AND NEW YORK

First published 2016 by Routledge

2 Park Square, Milton Park, Abingdon, Oxon OX14 4RN
711 Third Avenue, New York, NY 10017, USA

Routledge is an imprint of the Taylor & Francis Group, an informa business

First issued in paperback 2017

British Library Cataloguing in Publication Data
A catalogue record for this book is available from the British Library

Library of Congress Cataloging-in-Publication Data
European space policy : European integration and the final frontier / edited
by Thomas Hörber and Paul Stephenson.
    pages cm
    1. Astronautics and state–European Union Countries. I. Hörber,
    Thomas, editor. II. Stephenson, Paul (Paul J.) editor.
    TL789.8.E9E878 2016
    629.4094–dc23   2015012305

ISBN: 978-1-138-02550-9 (hbk)
ISBN: 978-1-138-03903-2 (pbk)

Typeset in Times New Roman
by Wearset Ltd, Boldon, Tyne and Wear

Contents

Figures

Tables

Contributors

Ulrich Adam is Secretary-General of the European Agricultural Machinery Industry Association (CEMA), Brussels, Belgium.

Angela Carpenter is Visiting Researcher at the School of Earth and Environment, University of Leeds, UK.

Jakob Feyerer is a PhD candidate at the Department of Political Science, University of Vienna, Austria.

Antonella Forganni is Research Associate at the EU–Asia Institute, ESSCA–School of Management, Angers, France.

Christina Giannopapa, PhD, is in charge of Relations with Member States in the Director-General's Cabinet, European Space Agency HQ, Paris, France.

Rik Hansen, is a PhD candidate at the Leuven Centre for Global Governance Studies, Leuven University, Belgium.

Thomas Hörber is Full Professor at the EU–Asia Institute, ESSCA – School of Management, Angers, France.

Harald Köpping Athanasopoulos is a PhD candidate at the University of Liverpool Management School, UK.

Veronica La Regina is Space Expert at the EU–Japan Centre for Industrial Cooperation, Tokyo, Japan.

Lucia Marta is a consultant in European space policy, R&D and programs, and is a Research Fellow at the Foundation pour la Recherche Stratégique, Paris, France.

Iraklis Oikonomou is a political scientist and holds a PhD in international politics from the University of Wales Aberystwyth, UK.

Emmanuel Sigalas is Associate Research Fellow at the Institute of International Relations (Prague, Czech Republic). Previously he was visiting professor at Carleton University (Canada) and Assistant Professor at the Institute for Advanced Studies (Vienna, Austria).

Paul Stephenson is Assistant Professor at the Department of Political Science, Faculty of Arts and Social Sciences, Maastricht University, the Netherlands.

Jan Wouters is Full Professor of International Law, Jean Monnet Chair and Director of the Leuven Centre for Global Governance Studies and Institute for International Law, Leuven University, Belgium.

Foreword

The EU is currently experiencing a multi-faceted crisis, with low growth rates, high levels of unemployment and increasing poverty in some regions. This crisis has reduced public support for the EU, as many people no longer see European integration as a solution, but rather as the root of the problems. It therefore needs a certain amount of courage and dedication to make the case for a European space policy and to advocate its role in the ongoing process of European integration, as is attempted in this book. European space policy is not an important factor in winning public support or national election campaigns. It is not at the heart of present EU policy-making. Its development does not follow election schedules, but rather 'out-of-the-box' thinking, based on a long-term-vision that space matters and that the peaceful exploration of space is one of the most crucial future challenges faced by humankind.

Without such thinking, we, as Europeans, would never have achieved what are now considered to be lighthouse projects, such as the establishment of our independent 'cosmic ear-and-eye' systems, Galileo and Copernicus, or the exceptional success of the Rosetta mission and the important European contribution to the ongoing mission of the International Space Station. It has helped to develop a strong and competitive position in the development and launching of satellites. However, in general, and for much too long, space policy has been something for the experts, dedicated politicians and space enthusiasts. Space has not yet become a popular subject, or a policy, that the majority of European citizens would consider as being crucial for our common future.

A recent special Eurobarometer report emphasised that the general public still underestimates the role of space. Job creation and health issues were seen as the most important EU priorities for the next 20 years, whereas space was clearly not seen as an immediate priority. However, the majority of EU citizens strongly support the commercial and public exploitation of some space activities, with the highest support rates for space technologies where the benefit is felt directly and immediately. A total of 74 per cent of those surveyed indicated that they thought space activities were beneficial for increasing our knowledge of the weather, 73 per cent underlined the importance of a better understanding of climate change and 72 per cent supported a positive impact on environmental protection and more efficient agricultural production. A majority of 57 per cent also supported

the view that space activities can lead to medical progress and 58 per cent associated space policy with job creation. However, only 37 per cent endorsed the
statement that space activities will improve security in the EU. When it comes to
space exploration, the result was extremely mixed, with 46 per cent supporting
further investment in space exploration, but 47 per cent opposed to this. In total,
29 per cent of those interviewed saw space exploration as a driver for new technologies and 24 per cent as a means of making scientific discoveries; however,
more than 50 per cent expressed interest in becoming better informed about
European space activities and policies.

From my point of view, and based on my experience as the member in charge
of space policy on the Barroso-1 Commission, I concluded from these findings
that the development of a European space policy needs both leadership and communication. Only if EU leaders are more outspoken and supportive of stronger
European investments in space will it be possible to mobilise the considerable
public investment required to drive independent European space activities.
Public opinion needs to embrace the idea that the EU requires more of its own
independent resources in space to master the challenges ahead. Under the present
circumstances, this is a very difficult, but not impossible, task. As the authors of
this book rightly emphasise, space policy is a strategic long-term choice.

Shortly after his inauguration as President of the European Commission, Jean-
Claude Juncker raised the concern that the EU is not respected as a serious global
player. He saw notably major deficiencies in foreign and security policies and suggested the creation of an EU army. Although I fully agree with the analysis of
Juncker that the EU risks further marginalisation in international relations, he
would probably have been well advised to also raise the issue of a well-coordinated
and solidly funded European space policy if the EU wants to defend its strategic
goals and interests in the emerging world of multiple space powers. Mastering
space is a precondition for the survival and defence of the European way of life as
it is a tool to ensure the peaceful exploration of the universe.

Innovation remains at the heart of global competitiveness, ensuring peace and
security. Front-runners in new technologies will benefit from first-mover advantages and will be able to commercialise their investments in term of welfare generation. They will be able to ensure their own security interests and will not be
dependent on the choices of others. They may contribute to an informed and
transparent global debate on what is required to prevent wars and fight terrorism,
to preserve the environment and to cope with the challenges of climate change.
Space activities are an indispensable tool to this end. If we are not innovative
enough in this area, we will become political bystanders and economically mar-
ginalised. Job creation will not happen in the EU, but somewhere else. Intelligence about international crises or natural disasters will not be available in the
EU, but somewhere else. We will not be able to convince partners about what is
needed to fight poverty or climate change based on own knowledge, but only
based on knowledge from external sources. For all these reasons, I fully support
the theses of this book that the EU should develop a step-by-step approach
towards a truly European space policy.

Until now, European ambitions in space have been rather limited compared with the engagement of the USA. India and China are now catching up rapidly with the USA and Russia, having fully appreciated that space policies are essential to their future. Therefore the EU will need to define a level of ambition and deal with a number of complex issues. One difficulty will be convincing those member states who do not have strong capacities in space research and space utilisation that there is an added value for the Community. It might also prove difficult to convince the small number of leading space powers in the EU that they should share the benefits of space research and space activities with the remaining Member States. From a purely institutional point of view, space policy is an EU responsibility, but it lacks sufficient funds and does not have sufficient expertise in leading concrete projects. This expertise is concentrated in the European Space Agency, which is intergovernmental in its structure and goes beyond the EU. Furthermore, European Space Agency activities are based on the principle of *juste retour*, which is not applied in the context of EU integration.

It will therefore need a lot of good will and political courage on both sides and a common vision about the challenges to make substantial progress under the present framework conditions. This is, however, the only way ahead, as the framework conditions cannot be changed immediately. Full Europeanisation of space policy and space activities is the next logical step, but we need to remain realistic about the time frame required for such a leap.

Finally, and as part of the long-term vision required for space, we need a strong consensus about space exploration or, more precisely, the exploration of our solar system. The USA has started to prepare a mission to Mars, underpinning my conviction that the desire to know and understand that which exists in our neighbourhood in space and to better understand our place in the universe is deeply rooted in human nature. The technical hurdles to achieve such a long-term mission are very high and the innovations required will put a strain on many budgets, but the benefits are immeasurable, both in terms of scientific and commercial value.

For Europeans at the current time, however, I see independent European access to space as an essential requirement. We must have the technology to bring our people back and forth to the International Space Station, still the most important symbol that space exploration is a common endeavour of the whole human race. Such common projects, based on an equal contribution of all the space powers, are the best prevention of a race to space. Although no one can be sure whether the race to space has already started, it is important for us to have the means to bind nations together rather allowing them to fall apart.

Would a European space policy have the power to reach out to the hearts of European citizens, to create a new bond to unite us? This is one of the driving questions of the book and is very difficult to answer. There will always be dreamers, and only centuries later will it become clear whether the dreamers have opened the door towards progress, or whether they just had a nice dream. Space policy has the potential to foster what European integration has promised: peace and prosperity. It will always have the potential to attract the minds and hearts of many. The more success stories are written, the more people will appreciate space activities as a way to uncover the uncountable secrets that nature and the universe still hide. As long as humans exist, people will watch the stars with fascination. However, those who have seen the earth from space have all come back with a similar experience: a much deeper understanding of how precious and fragile our home planet is, and with a stronger desire to protect our home planet, the earth. If European space policy contributes to that end, it would benefit us all.

I am most grateful that this book offers insights, experiences and facts to trigger a new debate on European integration and European space policy.

Günter Verheugen
Former Vice-President of the European Commission

Preface

Thomas Hörber has a keen interest in all things space, including science fiction. Paul Stephenson has previously written about transport infrastructure and 'fell into space' via Galileo. Both came together at the first of five space policy workshops at the École Supérieure des Sciences Commerciales d'Angers School of Management, along with about 30 researchers, both academics and practitioners.

What became quickly apparent was the huge level of interest in space policy from scholars working in various policy domains and in individual EU institutions. The workshops brought together political scientists, historians, economists and lawyers, as well as people working 'full-time in space' at various agencies.

'Space' seemed to be an expansive, cross-cutting, umbrella issue that was somehow relevant to many EU policy areas, but was also the key to the transnational challenges Europe was facing: environmental protection, job creation, security and defence, migration, mapping and the Single Market. It was an area that was cutting edge in terms of showcasing European technological prowess, as well as visionary policy-making.

The paradox was that there is a significant community of space scholars worldwide – as exemplified through the academic output in leading journals such as *Space Policy* – and yet the European political science community did not yet seem properly mobilised, despite panels on European space policy at the University Association for Contemporary European Studies annual conference for the past four years. This might be because space had an image problem and is seen as highly technical, very far-off (literally) and, well, a bit nerdy.

From the enthusiasm of those first workshops and the incredible variety of papers delivered, both in Paris and at conferences in Cambridge, Passau, Bruges, Leeds and Cork, it seemed sensible to try to bring space research together in a book. And it had to be the University Association for Contemporary European Studies/Routledge series because we wanted the chapters to be read by the broader European studies community, not technical specialists.

We wanted to show the relevance of space policy to everyone. We sought to 'make a case for space' and, in so doing, to communicate its great potential for European integration.

Thomas Hörber
Paul Stephenson

Acknowledgements

The editors would like to thank all the contributors, who have done an excellent job in bringing this edited book together. Through their dedication, we hope to have achieved a good introduction of space policy into European studies. In this context we would also like to thank the University Association for Contemporary European Studies for their support of our work, through space panels at the annual conferences, as well as funding at key points during the development of this space policy research project. We would also like to thank the École Supérieure des Sciences Commerciales d'Angers (ESSCA) for their continued support in organising the annual ESSCA European Space Policy Workshop. Their funding and encouragement of this research, particularly by the head of the EU–Asia Institute at ESSCA, Professor Albrecht Sonntag, have made this book possible. We thank the reviewers of our initial book proposal, whose recommendations helped us to strengthen the focus and scope of the book. We also thank Elissaveta Radulova, whose own use of framing in European social policy inspired Paul Stephenson in his original approach to using framing as a tool to analyse Galileo in his *Space Policy* article, which we eventually used as a guiding principle and common lens for analysis throughout these chapters.

Thomas Hörber
Paul Stephenson

Abbreviations

ASD	Aerospace and Defence Industries Association of Europe
BeiDou	Chinese Satellite Navigation System
CAP	Common Agricultural Policy
CJEU	Court of Justice of the European Union
COPUOS	Committee on the Peaceful Uses of Outer Space
CTP	Common Transport Policy
DBFO	Design Build Finance Operate
DG-MAF	Directorate-General for Maritime Affairs and Fisheries
ECJ	European Court of Justice*
EDA	European Defence Agency
EEAS	European External Action Service
EGNOS	European Geostationary Navigation Overlay Service
ELDO	European Launcher Development Organisation
EMF	European Metalworkers' Federation
EMSA	European Maritime Safety Agency
ENVISAT	Environmental Satellite
ESA	European Space Agency
ESDP	European Security and Defence Policy
ESP	European Space Policy
ESRO	European Space Research Organisation
EU	European Union
EUMETSAT	European Organisation for the Exploitation of Meteorological Satellites
FAO	United Nations Food and Agriculture Organisation
FOC	Full Operational Capacity
GDP	Gross Domestic Product
GEOSS	Global Earth Observation System of Systems
GJU	Galileo Joint Undertaking
GLONASS	Russian satellite navigation system
GMES	Global Monitoring for Environment and Security
GNSS	Global Navigation Satellite System
GPS	Global Positioning System
GSA	European Global Navigation Satellite Systems Agency

IFOAM EU	European Umbrella Organisation for Organic Food and Farming
IFPRI	International Food Policy Research Institute
industriALL	Industrial Global Union (mining, engineering, manufacturing)
JO	*Journal Officiel/Official Journal*
MFF	Multiannual Financial Framework
NASA	National Aeronautics and Space Administration
OECD	Organisation for Economic Development and Cooperation
OHB	Orbitale Hochtechnologie Bremen
PPP	Public–Private Partnership
Roscosmos	Russian Federal Space Agency
SCPI	Sustainable Crop Production Intensification
SOLAS	International Convention on Safety of Life at Sea
TEC	Treaty establishing the European Community
TEN-T	Trans-European Transport Network
TENs	Trans-European Networks
UN	United Nations
UNCOPUOS	United Nations Committee on the Peaceful Uses of Outer Space
UNODC	United Nations Office on Drugs and Crime
USSR	Union of Soviet Socialist Republics
WEU	Western European Union

---

* Throughout the book we use the common term 'European Court of Justice' or its abbreviation, the ECJ, in reference to the Luxembourg institution. However, in Chapter 7, the author uses the term 'Court of Justice of the European Union' or CJEU, which is now increasingly used since the Lisbon Treaty to refer to the whole institution comprising three courts, one of which is the Court of Justice (in addition to the General Court – formerly the Court of First Instance – and the Civil Service Tribunal).

Introduction

Towards a European space policy?[1]

Thomas Hörber

Genesis and aim of this edited book

The aim of this book is to evaluate the state of European integration at the current time and the part that European space policy could play in furthering this process. As such, we consider European space policy as a tool for the analysis of European integration, rather than an object of study in itself. Space policy is presented in this book as an original and timely lens for studying developments in European policy-making. Space policy has become a fascinating object of study, involving multiple and diverse actors and interests in the European integration process. The contributors to this book look closely at most of the major institutions of the EU, as well as other actors, in their analyses of the dynamics of the European policy process.

Let us think back to the carnage of the Second World War and the momentous leap of faith that Robert Schuman and Jean Monnet took, together with other leading figures, including Konrad Adenauer, Paul-Henri Spaak and Alcide de Gasperi. Much has been achieved in recent decades. Yet, after 60 years of integration, the next generation of the general public continues to ask: 'What does the EU do for me?' and 'What does the EU stand for?'. The authors in this book all see space policy as an important contributor to – and catalyst for – the European integration process. A coherent European space policy could provide an answer to questions about the purpose of European integration. Might space be a policy solution to the bigger challenges of identity and legitimacy in Europe? This in itself is a thought that casts doubt on the current status quo and may offer hope beyond the current malaise of the EU. In that sense, this book does not seek to be an all-encompassing objective analysis of European integration, but instead is concerned with policy analysis.

The authors analyse European integration – without neglecting its shortcomings – and seek to develop ideas about future strategies for fostering that process, notably through a European space policy. Space is very much at the cutting edge of current EU policy developments. Recent phenomena, such as climate change, rapid developments in crisis management and prevention, and fast-paced transformations in our natural environment, all need space capabilities to 'manage' them. Against this background, there is an element of advocacy in the very remit of this book, which is conveyed by the authors' conviction of the rationale and potential of such a strategy for the benefit of Europe. To the best of our capabilities, we have tried not to become carried away by our enthusiasm for the topic and, as good academics, have injected methods and instruments of academic objectivity, as explained in the following section on methodology. The editors hope, however, that such subjectivity is complemented by sound analysis and evidence-based argument, making the contents an enjoyable read for those willing to embark with us on an (intellectual) journey to the stars.

Framing theory

The concept of framing acts as a unifying bond for the analysis throughout the chapters. Policy frames have been widely used in the European studies literature (Schön und Rein 1994; Kohler-Koch 1997; Mörth 2000). It is often argued in the framing literature that crises trigger reform as a result of the breakdown or the confirmation of policy frames (Boin *et al.* 2009: 81). A crisis 'typically generate[s] a contest between frames and counter-frames concerning the nature and severity of a crisis, its causes, the responsibility for its occurrence or escalation, and the implications for the future' (Boin *et al.* 2009: 82; Daviter 2012: 3; Jones and Baumgartner 2002: 298). Crises can be a catalyst for change, but the crisis must be articulated through language: messages, stories, accounts and other narrative devices. Kohler-Koch (2000: 516) asserts that framing theory pushes the limits of our theoretical understanding of the EU a bit further because it creates a new way of understanding what is beyond the nation state – that is, our current mental map. Frames establish new norms and rules that make the political 'reality' beyond the nation state comprehensible –in fact, frames conceive of and project their own realities about what has occurred and what needs to be done. This applies to the whole policy spectrum, where frames help to shape political life and, in consequence, policy priorities.

The early mental map (Surel 2000: 498) of European space policy was French, especially in the 1960s and 1970s. Modifications were incorporated thanks to other space-faring nations, such as the UK and Germany. However, in the aftermath of the creation of the European Space Agency (ESA) Convention, space ventures became increasingly European, making room for other nations, such as Italy and Belgium, to develop their own space initiatives alongside the lead nation, France.

The EU itself has been taking an increasing interest in space. The breakdown of the Galileo public–private partnership (see Chapter 13) and the consequent Europeanisation of the global positioning system is a good example. The EU considered the Galileo satellite navigation system too important to let it collapse and therefore reshaped its legitimising frames to secure agreement from inside – the Commission and the European Parliament – as well as from the Member States/Council. It appealed to the general public through a substantial public relations campaign (Boin *et al.* 2009: 83–84). Arguably, the Commission gained power by securing the financing of a costly programme through the EU budget

(Mörth 2000: 174). The European Parliament also increased its influence through its support of Galileo and the space project (see Chapter 4). Thus we saw a shift of power towards the supporters of Galileo, which partly explains why the Commission and the European Parliament supported the project, as predicted in the power dynamics model of Boin *et al.* (2009: 88). In such a setting, frames can be used for the analysis of the power struggle between European institutions, within them – for example, in different Commission Directorate-Generals (Mörth 2000: 174) – and between the EU and the Member States. This means taking into account multiple actors and institutional dynamics, best conceptualised in neoinstitutionalism (March 1994, 1997; March and Olsen 1996). Long-term historical policy studies, which Dudley and Richardson (1999) see as the only way to understand the development of frames, have only recently been taken on board in publications on space policy. As a result, space policy is arguably not yet sufficiently understood.

European space policy has always been based on the fundamental principle of peaceful use (see the ESA Convention, available from the ESA web site). However, dual-use technologies are part and parcel of space technologies, despite the fact that space is generally defined as a 'civilian' policy –for example, Galileo and even more so the ESA. The frame of dual-use has allowed military funding into civilian space programmes by maintaining space infrastructure for 'peaceful purposes'. Their design remains civil, but security is hardened up and therefore also potentially viable for military use. For the EU, this tendency has also become increasingly common place, for example in research funding since the Fifth Framework Programme. Mörth (2000: 178) calls this a spin-in effect, in which we are not so much seeing a militarisation of space civilian technologies, but rather an increasing dependence of the military on these civilian technologies. The closer relationship between the Western EU (WEU) in becoming the defence agency of the EU is a good example of where a formerly military organisation acquires civilian political institutions that become the decision-making bodies. The WEU created the EU satellite centre, which then became an EU agency. The WEU does not have military satellites, but purchases images from military sources for intelligence dossiers in support of the European Security and Defence Policy. Such tendencies illustrate the EU drive for greater independence on the international stage by acquiring defensive capabilities (Mörth 2000: 179). That falls in line with the founding ideal of peace (Hörber 2012: 78) insofar as dual-use technologies do not serve aggressive purposes, but could defend European interests in a potentially dangerous international environment.

Frames in the space sector enable the 'sense-making' (Mörth 2000: 186) of very complex policies, which is necessary even for the decision-makers themselves (Baumgartner and Mahoney 2008: 440). At the same time, it seems that entertaining several frames, as the Commission in particular does, acts as a way to capture different constituencies with different agendas (Jabko 2006; Baumgartner and Mahoney 2008: 444). This is normal because no policy is one-dimensional (Baumgartner and Mahoney 2008: 436–437). Accounting for exactly this complexity of policies requires multiple actor perspectives. Daviter

(2012) has even argued that the existence of different frames is the reason why decision-making is possible, because they provide us with choice (Daviter 2012: 8). This also reflects very well the different world views existing in our societies. Thus frames, and particularly several frames at the same time, maintain the vagueness of (space) policies (Mörth 2000: 186). Here we are back at the roots of framing theory in discourse theory, which postulates that discourse connects different issues in a sufficiently 'sutured' way (Laclau and Mouffe 1985). The key objective is to keep the discourse coming so that by participating in it – and, indeed, manufacturing it – different parties will develop a loyalty towards the system, whether it is the EU or the ESA. Framing theory is particularly suited to understanding the changing relationship between these two organisations, which itself will define European space policy and even European integration.

We use framing theory in a loose way in this book – that is, without imposing a specific frame on authors, because we did not want to dilute the rich empirical findings of each chapter by forcing a rigid theoretical framework upon them, or requiring them to look for 'a certain way of talking'. What we sought the authors to do was to identify the particular frames operating in the discourse of different actors. As such, in each chapter, the authors may identify one or more operating frames. These vary from the legal frames operated by the Court of Justice of the European Union and applied in fledgling case law on space-related matters, to the high politics frames that the EU uses in its international relations. What framing theory has been used to do in this book is to examine actor and institutional interests, to provide the reader with an explanation of how European space policy operates through a discursive lens. For the moment, this overview of framing theory and how it has been used in this book will suffice. A more detailed description of the analytical merits of this approach is provided in Chapter 2.

Guiding ideals of European integration

What were the ideals driving the early process of European integration? The realist perspective on the Europe of the 1950s might suggest 'simply' peace and prosperity. Arguably, other ideals were at play, other visions for the continent to engage in a collective political and societal endeavour. Arguably, these ideals have been lost, clouded by institutional navel-gazing, media concerns of an inflated bureaucracy, systemic failings and accusations of democratic deficit. In this vein, we outline why a European space policy could give the European integration process new direction and purpose. The following section explains the context of the founding ideals of European integration. It follows the early idealist discourse, but with reference to very concrete integration efforts, such as the Schuman Plan (see Lipgens 1982; Monnet 1976; Lejeune 2000). Such an idealistic perspective is a deliberate choice because, until now, the embodiment of the European integration process, the EU, has been compared with nation states as well as with international institutions. But are we in the habit of comparing the EU to nation states? Is this really warranted? The idealist discourse is one of several that reneges on this classical comparison, but rather argues that European integration was something new. Particularly with regard to the nation state, the EU has a different function to fulfil. It is argued in this introduction and throughout the book that the EU has and should fulfil an idealistic leadership function. What does that mean? Very simply, the argument is that the EU has led us to where we are today – that is, to a prosperous and peaceful Europe. Not everyone is convinced of that reality. Criticism of the European integration process is widespread and growing, as recent European elections have shown. A longer term vision of European integration might need political conviction and financial commitment beyond the economic policy cycle and the short-termism of elected politicians.

All is not well in Europe. Why this is so may, in fact, be due to the success of the European integration process. The early integration ideals of peace and prosperity have been fulfilled and they have served their pioneering function. As such, they are no longer sufficient to act as a driving force for further European integration (Hörber 2006, 2014; Jabko 2006; Parsons 2006). A European space policy might have the potential to renew the idealistic forces of early post-war integration and make twenty-first century Europeans reach for new horizons.

EU challenges

Over the past 60 years and for a variety of reasons, there have been periods of faster and slower European integration. Dominant personalities such as Robert Schuman or Charles de Gaulle have played defining roles. Political events such as the end of the Cold War and German reunification – closely linked to the Maastricht Treaty, which took us from the European Community to the EU – have been key factors in European integration.

Pro-integrationists have at times been frustrated by the slow speed of integration (Monnet 1976). Such criticism in itself might not be very significant, because such crossfire has been a fact of life in European integration ever since it was born. However, it will be argued here that a European space policy can positively contribute to the European integration process through fostering a common European identity in space endeavours.

Nevertheless, the idealistic basis of the European integration project is also open to criticism. We said that the ideals of peace and prosperity, which defined the direction of early European integration, have been more or less fulfilled. However, the current conflict in Ukraine and poverty in some Eastern European countries (and, for different reasons, in Western European countries), are the most obvious signs that a level playing field has not yet been achieved through market integration. Space infrastructure will secure integration and development and has a role to play in crisis management. However, the continued existence of war and poverty in Europe today bears no comparison with the Europe of the immediate post-war period. Moreover, within the EU, such threats are virtually extinct. The EU is working actively by way of diplomacy and ground missions to tackle the economic and military threats in neighbouring Eastern European countries. Examples are the economic development programmes, neighbourhood partnership agreements and the controversial engagement of the EU in the Ukraine crisis in 2014.

These pioneering ideals of the European integration process – peace and prosperity – have been largely fulfilled and, as a consequence, they no longer have the potential to pull the integration process forward. These ideals no longer drive governments or people, although arguably these ideals still hold sway in the New Member States that have joined since 2004.

Towards new ideals

This book argues that the European integration process needs a new direction, perhaps even new ideals. Clearly, peace and prosperity must be sustained, but the EU has generally proved that it can do this, at least within the EU itself. Outside the EU, a notable exception to peace was the Balkans wars of the 1990s. What if we now went beyond these founding ideals, while keeping the spirit of early European integration? The founding fathers translated these far-fetched ideals into concrete policies, which aligned the European nations, setting out visionary common objectives for the future. We might do the same by proposing a European space policy, which could provide an innovative agenda to 'shape' European politics and society. Indeed, there are indications that space is already playing a major role. There is now a new Commissioner for Transport and Space, Maroš Šefčovič. The existence of a portfolio for space is an indication that it has become a high profile issue. Space policy has become mainstream; it is contemporary, fast-moving and has practical applications. No longer is space seen as a technical issue area on the margins of Community policy. The ESA Convention (1975) celebrates its fortieth anniversary this year. Over four decades it has built up impressive technical expertise. In the following chapters, we show how space has matured as a policy field, now formally recognised across the EU institutions.

Is space an insurmountable barrier to integration, or is it a barrier we can remove? Either we shirk the challenge of common space exploration, or we recognise its economic and political potential and forge an action plan. Institutionally, such a venture clearly means a much more prominent role for the ESA. Following Jean Monnet, whose conviction was that no daring project can succeed under unanimity rule (Monnet 1976: 333, 413), it seems wise to bring the ESA fully within the supranational Treaty framework of the EU with qualified majority voting (now called the Ordinary Voting Procedure), particularly in the Council of Ministers. The joint Space Council between the EU and the ESA was a first step in this direction (see Chapter 5).

Beyond the institutional arrangements, there are huge questions over the budget. Compared with NASA, the ESA annual budget of €3.34 billion – of which €623.9 million comes from the EU – is relatively small. If the ESA is to take up a leadership role, the only reasonable way of achieving this may be to increase its spending capacity. A gradual increase dedicated to specific projects, such as the Galileo global positioning system, has worked well in the past, despite the experience of public–private partnerships (see Chapter 13). The ESA's space administration has also developed considerable expertise in managing space programmes. Matters of finance, leadership and programme management will need to be negotiated further. The following chapters contain some ideas as to the type of initiatives we may see in the future. Exploration as a new guiding ideal for the EU could effectively harness the very human driving instinct to find out more about who, what and where we are as Europeans. Venturing into space – the final frontier of integration? – arguably appeals to the boundless curiosity of the human spirit.

The methodology section (Part I) begins with a chapter by Thomas Hörber, which outlines the development of the ESA from its predecessor organisation with the frame of independence for Europe, promoting technological innovation while accepting that budgetary constraints were much tighter in Europe than in the USA. Paul Stephenson continues with a chapter on framing as a tool for analysing European space policy. It contains an in-depth overview of framing theory, making clear its applicability to European space policy, and this edited book as a whole.

Part II, on space polity and institutional perspectives, begins with a contemporary analysis of the relations between the ESA and the EU. Chapter 3 outlines the discourse leading to a much closer relation between the ESA and the EU from the 2000s onwards. Thomas Hörber highlights the growing political importance of space in European politics, as shown through increasing budgets, as mentioned for Galileo, or institutionally with the establishment of the ESA–EU Space Council as a joint session with the Ministerial Councils. The main frame behind this tendency is democratic legitimacy. This does not mean that power politics do not interfere in such matters; the Commission has had misgivings about the Space Council since the early 2000s.

Chapter 4 identifies how the European Parliament has been engaged in European space policy and the way its own frames have emerged. This chapter investigates how the Europeanisation of space has been portrayed as a legitimate development within the European polity. It shows convincingly that the European Parliament has long supported space policies. Space is conveyed as a highly promising policy field from which Europe should reap benefits, including the principal objective of European independence. The analysis shows that the European Parliament has acted strategically by adapting the frames through which it justifies space over time, to ensure that its reasoning remains in line with the EU's varying strategic priorities.

Chapter 5 examines the Council of Ministers and its role in setting the frames around European space policy. Rather than emphasising the Council's dominance as a legislator, the chapter argues that the Council has instead acted as the primary frame selector in the process of justifying the EU's ambitions to engage in space. In this context, the chapter concludes that human activity in space is increasingly seen as a means to some material end, rather than as an end in itself. Particular attention is given to an idealistic frame in European space policy, which asserts that priority should be given to space exploration as it provides us with immeasurable insight into the place and purpose of humanity in the universe.

Chapter 6 looks at the inner workings of the Commission, including its framing of the Galileo venture. This chapter traces the way in which the European Commission has framed and reframed the issue of EU satellite navigation in recent years. It investigates how the agenda-setter has 'talked about' space policy, with a particular focus on Galileo, examining how the Commission's own institutional discourse has evolved over a decade through the use of 'frame sets'. In so doing, it shows how the Commission has chosen to present the issue as politically and economically desirable for the EU. The earlier methodology chapter by Stephenson (Chapter 2) also looks close up at the language and vocabulary of EU legislation on Galileo.

Chapter 7 looks at the role of the Court of Justice of the European Union in European space policy. Unsurprisingly, there was no concrete case to be found, because space policy only became a EU competence with the 2009 Lisbon Treaty. However, a number of Lisbon Treaty rulings are identified in the field of aeronautics and telecommunications, which are likely to develop into reference cases for future decisions on European space policies. In previous decades, the Court of Justice of the European Union has often acted as a federator for the EU, and particularly the internal market, pushing the frames of free movement of services and goods. These areas are highly relevant to space policy in the EU. It will be revealing to see whether the Court of Justice of the European Union develops a similar pro-integration dynamic, with space policy becoming central to the functioning of the internal market as, arguably, it will if the increase in e-commerce and e-services continues.

In Part III, the politics section, we begin our internal perspective on the EU with a focus on industrial interests with the example of the agricultural sector (Chapter 8). The Common Agricultural Policy is still the most important spending line of the EU budget, even if it appears disguised as other things, such as landscape management and territorial cohesion. Space, through technology, has been essential for the modernisation of the agricultural sector. Finding widely accepted technology solutions that can raise productivity levels and promote environmental protection in farming has proved to be particularly challenging. As it can achieve both sustainability and productivity growth in agriculture, farming by satellite – commonly termed precision agriculture – has emerged as a strategically important approach that can help to meet the challenge of feeding a growing world population. EU space policy, and therein the rapid growth of satellite technology applications in farm machinery, presents the agricultural machinery industry with a unique opportunity to transform adverse and outdated views about agriculture and farm machinery. The agricultural industry has used space technologies to reframe its image and outlook in EU policy-making.

The great European infrastructure projects of the trans-European networks are identified in Chapter 9 as the Commission's agenda to complete the Single Market in 1992 and to increase the competitiveness of the EU in the world. The trans-European networks are shown to be one of the main tools for managing the progress of the EU and space is becoming increasingly important in this context. Thus the frame of infrastructure development was extended to including space technologies and the consequent institutional connections to the ESA. Trans-European networks were justified using three different frame sets. Their rapid development, against the backdrop of 30 years of neglect in European transport policy, arguably helped to create a policy-making environment and to secure an ideational consensus that would subsequently allow the development of Galileo. Transport policy can thus be seen, through comparative historical analysis, to be complementary to space policy, while also proving a policy foundation for common large-scale infrastructure investment and long-term planning.

Chapter 10 examines trade union interests in European space policy with the example of the European Metalworkers Federation. In doing so, past framing activity is considered very critically. Mörth (2000) introduced framing in the study of EU defence policy integration, whereas Stephenson (2012) inserted the concept into the field of EU space policy and the case of the Galileo project. Undoubtedly, the notion of framing has yielded fruitful results in terms of both theoretical and empirical output, shifting the emphasis from a supposed objectivity of interests to the linkages between interests, ideas and institutions that are inherently subjective. Nevertheless, it is asserted that there is an inherent problem with this approach, namely the lack of a theory of power embedded in it, or, in other words, the ontological primacy of discourse and ideas over material interests. Chapter 10 provides an analysis of the discursive patterns articulated by the European Metalworkers' Federation in the interrelated policy fields of space and defence. It argues that framing cannot account for discursive regularities, nor for the maintenance of a stable, pro-industrial stance at the core of the rhetoric of European trade union representatives. Such a set of patterns is the outcome of the exercise of hegemony by the ruling social and economic forces of the internationalised aerospace industry. We might expect to see the multiplicity of interests, discourses and narratives flourishing in the European Metalworkers' Federation, but the empirical data show a dominant pro-industry discourse among its ranks. This chapter posits that the defence and space industry exercises an ideological dominance to which trade unions submit – rather than the trade unions designing their own frames or making their own interpretation of them.

In international politics, the growing assertion of the EU as an actor also features European space policy as a tool. What is internal and what is external politics is hard to define in the EU context. Chapter 11 examines external politics from the perspective of the nation states. The main objective of this chapter is to provide an up-to-date overview of the national space governance structures, strategic priorities and industrial capabilities in the ESA Member States. This approach has its merits, because the ESA has 20 Member States, all with different motivations, expectations and capabilities. These factors shape their industrial structure and vice versa. A number of countries engage in space activities exclusively though the ESA, whereas others also have their own national space programme.

Chapter 12 takes a true international relations perspective on space policy, arguing that space is at the heart of Europe's international relations with the rest of the world, involving the Member States of the ESA and the EU. As such, analysing the framing of space policy is complex because of multiple, often competing, internal and external political interests and values. This chapter identifies the dynamics of the framing process to better understand the weaknesses of EU institutional reforms. Different framing policy options are expressions of different interests – be they market- or defence-oriented – by advocacy coalitions and individual policy entrepreneurs. The chapter examines different discursive approaches to the EU's international cooperation with third countries and/or overseas regional blocs. It identifies two main frames: geopolitics and industrial policy. The first is typically adopted in a political context when dealing with countries active in space exploration, where EU cooperation ranges from mere coordination actions to sharing and managing space-based assets. The second frame operates when cooperating with third countries without space-based technologies, where the EU attempts to promote its own industrial interests to gain a greater share of overseas markets.

In the final policy section of this book, policies directly relevant to space – and their implementation – are analysed. Chapter 15 looks in detail at the process of developing the Galileo satellite navigation system. Galileo was first framed as a promising public–private partnership, based on the rationale that such a venture would save public money. After severe setbacks, the public–private partnership collapsed in 2007. The Commission's framing of the project then had to change mainly to one of European independence from the American global positioning system, but also to one of technological innovations, underlining all the predicted positive spin-offs, such as cutting-edge technology, increases in productivity and consequent employment. This chapter also provides an analysis public–private partnerships and how they might be adapted to future space projects.

Chapter 14 examines EU industrial policy for the space sector. The rise of the EU as a space actor, especially as an initiator, owner and operator of large-scale programmes such as Galileo and Copernicus, has raised a number of questions with regard to industrial policy. Based on the experiences of the Galileo procurement process and on present discussions on space industrial policy within the EU, this chapter finds that the EU's political ambitions in space are reasonably well-defined, but that the specific policy tools and legal instruments to put them into practice are far from complete. The authors analyse the frames used in the discourse of the European Commission and the European Space Council on the topic and conclude that it is high time for a European industrial space policy.

Finally, the policy analysis of Galileo and Copernicus in Chapter 13 reveals how satellite technology offers many benefits to Europe's maritime regions and its marine environment, ranging from tracking shipping to monitoring pollution and enforcing border controls. This chapter makes a wholly convincing case for space, showing which prominent EU policies need space technologies to ensure their proper functioning. Space technologies are transversal to many EU policies, which explains why space has found increasing interest from European authorities. Beyond the chapter, we might think of how satellite navigation can be used to monitor human flows during civil war conflicts, mass migration, forest fires, volcanic ash, air pollution and the melting of the ice caps.

The contributions in this book contain objective, evidence-based analyses. Their findings point towards the gradual development of a comprehensive, innovative and proactive European space policy. In the future, space may well become the answer to the questions 'What does Europe do for me?' and 'What does Europe stand for?' Market integration and conflict prevention need strong and secure space assets, including earth observation and monitoring capabilities for the management of climate change and the prevention of environmental disasters. Directly or indirectly, space is relevant to every EU policy area. This book hopes to place space at the heart of European studies, advocating that, through a European space policy, Europe might boldly go where no one has gone before!

Note

1 The gist of the introduction to this book has been published previously in two separate articles. The first was really a 'thought-piece' right at the beginning of my interest in European space policy: 'The nature of the beast: the past and future purpose of the European integration' (*L'Europe en formation*, 2006, vol. 1: 11–19). The second was a summary of the purpose of a space policy for the European integration process: 'World War II is over' as the Introduction to the Special Issue on European Space Policy (*Space Policy*, 2012, vol. 28, issue 2). Both lend themselves to giving a message to this book, outlining an objective for a future European space policy. This introduction will try give such a narrative of the usefulness of a European space policy.

References

Baumgartner, F.R. and Mahoney, C. (2008) 'The two faces of framing: individual-level framing and collective issue definition in the European Union'. *European Union Politics*, 9(3): 435–448.

Boin A., 't Hart, P. and McConnell, A. (2009) 'Crisis exploitation: political and policy impacts of framing contests'. *Journal of European Public Policy*, 16(1): 81–106.

Daviter, F. (2012) *Framing Biotechnology Policy in the European Union*. ARENA Working Paper 05/2012. Oslo: ARENA Centre for European Studies.

Dudley G. and Richardson, J. (1999) 'Competing advocacy coalitions and the process of frame reflection: a longitudinal analysis of EU steel policy'. *Journal of European Public Policy*, 6(2): 225–248.

Hörber, T. (2006) *The Foundations of Europe – European Integration Ideas in France, Germany and Britain in the 1950s*. Wiesbaden: Springer Science and Business Media.

Hörber, T. (2012) 'New horizons for Europe – a European studies perspective on space policy'. *Space Policy*, 28(2): 77–80.

Hörber, T. (2014) *War Experience, Changing Security Concepts and Research & Development in a Converging Post-war European Discourse*. New York, NY: Lexington.

Jabko, N. (2006) *Playing the Market – a Political Strategy for Uniting Europe 1985–2005*. Ithaca, NY: Cornell University Press.

Jones, B.D. and Baumgartner, F.R. (2002) 'Punctuations, ideas, and public policy'. In: F.R. Baumgartner and B.D. Jones (eds), *Policy Dynamics*. Chicago, IL: Chicago University Press, 293–306.

Kohler-Koch, B. (1997) 'Organized interests in European integration: the evolution of a new type of governance?' In: H. Wallace and A. Young (eds), *Participation and Policy Making in the European Union*. Oxford, Clarendon Press, 42–68.

Kohler-Koch, B. (2000) 'Framing: the bottleneck of constructing legitimate institutions', *Journal of European Public Policy*, 7(4): 513–531.

Laclau, E. and Mouffe, C. (1985) *Hegemony and Socialist Strategy*. London: Verso.

Lejeune, R. (2000) *Robert Schuman: Père de l'Europe*. Paris: Fayard.

Lipgens, W. (1982) *Die Anfänge der europäischen Einigungspolitik 1945–1950*. Oxford: Clarendon Press.

March, J. (1994) *A Primer on Decision Making: How Decisions Happen*. New York: Free Press.

March, J. (1997) 'Understanding how decisions happen in organizations'. In: Z. Shapira (ed.), *Organizational Decision Making*. Cambridge: Cambridge University Press, 9–34.

March, J. and Olsen, J. (1996) 'Institutional perspectives on political institutions'. *Governance*, 9(3): 247–65.

Monnet, J. (1976) *Mémoires*. Paris: Fayard.

Mörth, U. (2000) 'Competing frames in the European Commission – the case of the defence industry and equipment issue'. *Journal of European Public Policy*, 7(2): 73–189.

Parsons, C. (2006) *A Certain Idea of Europe*. Ithaca, NY: Cornell University Press.

Schön, D. and Rein, M. (1994) *Frame Reflection: Toward the Resolution of Intractable Policy Controversies*. New York, NY: Basic Books.

Stephenson, P.J. (2012) 'Talking space: the European Commission's changing frames in defining Galileo'. *Space Policy*, 28(2), 86–93.

Surel, Y. (2000) 'The role of cognitive and normative frames in policy-making'. *Journal of European Public Policy*, 7(4): 495–512.

Part I

Methodology

Setting the space agenda

1 Chaos or consolidation?

Post-war space policy in Europe

Thomas C. Hörber

Introduction

In the aftermath of the Second World War, space technologies became an important area of development for the leading world powers. The Sputnik shock in 1958 – when the Soviet Union proved that they could put a satellite into orbit – and the Moon landing achieved by the USA in 1969 were important technological milestones. They were also major events in the Cold War context, serious evidence that both sides had access to, and knew how to use, cutting-edge technology, strengthening the leadership status of both sides vis-à-vis their allies. Such endeavours required substantial resources that were not available to medium-sized powers such as France and the UK. These countries clearly understood that the formidable costs of a fully-fledged space programme could not be borne by any one of them alone. However, both the UK and France realised that such key technologies were indispensable, the reasons for which will be analysed in detail in the main text. Against this background, this chapter traces the progress of early bilateral cooperation, which culminated in the wider project of the European Space Agency (ESA).

Methodology

Framing theory has become increasingly popular in recent European studies. With its roots in discourse theory, it is sufficiently flexible and at the same time sufficiently sutured, as Laclau and Mouffe (1985) would say. Hall put it this way.

> ... politicians, officials, the spokesmen of social interests, and policy experts all operate within the terms of political discourse that are current in the nation at a given time, and the terms of political discourse generally have a specific configuration that lends representative legitimacy to some social interests more than others, delineates the accepted boundaries of state action, associates contemporary political developments with particular interpretations of national history, and defines the context in which many issues will be understood.
>
> (Hall 1993: 289)

The more fundamental contributions to framing theory come from the development of paradigm, as in Hall (1993, 1997, 1998), belief systems as developed by Sabatier (1993, 1998) and Kohler-Koch (2000: 514) and, to a lesser extent, from *référentiels*, as in Jobert and Muller (1987). More importantly, framing offers new insights into the often still vague politicking of the EU to the extent that we might assume that the theory was designed to explain the inner workings of the EU, where many frames are created to attract a constituency and where several of these frames coexist at the same time. They often come from different world views, e.g. neoliberalism versus Keynesianism in the 1970s (Hall 1992, 1993; Surel 2000: 497; Rein and Schön 1991: 264) or, more concretely, the French advocacy for independence in space technologies versus the UK's primary concerns for budgetary constraints, as will be shown in this chapter. As Surel asserts, '..cognitive and normative frames not only construct "mental maps" but also determine practices and behaviours' (Surel 2000: 498). More simply, 'a frame analysis helps to explain why actors want what they want' (Mörth 2000: 175). The political motivations for a consolidated European space effort will be the subject of this chapter

Analysis: politics

With the evident success of the US Apollo programme in the shape of the actual Moon landing, it would have seemed natural for Western European middle powers to join in the post-Apollo programme proposed by the USA, not least because of the substantial cost of space technology. The gist of the financial argument is well summarised by a British minister.

> There is no economic case for an independent European launcher. Launchers are the most expensive and least profitable items of space technology. Satellites, on the other hand, are relatively inexpensive and their applications in commercial ventures offer the best prospects of commercial return. The resources available in Europe are, by common agreement, small by comparison with those of the United States, and the Americans have an impregnable lead in launcher technology, while there is good prospect that Europe can compete successfully in satellites.
>
> (Onslow, *Weekly Hansard* 1973: 842)

However, scrutiny of primary sources – parliamentary debates either from the French *Journal Officiel* (JO) or the British *Hansard* or *Weekly Hansard* – shows that the first nodal point in the discourse is not partnership, but competition with the USA (McNair-Wilson, *Weekly Hansard* 1973: 828–829).

The post-Apollo programme: competing with the USA

Such competition, if it was meant to be serious, necessitated a comprehensive space programme on the European side, from space ports to launchers, satellite technology and even the capability to maintain humans in space. Naturally, both

the UK and France wanted to see their technologies used in any such comprehensive space programme. However, both also realised that an overarching
European programme was the key to successful implementation, as can be seen
in the following quotation.

> I find it difficult to see how Europe as an entity can operate a total space pro-
> gramme without having a total space capability.... This means that Europe will
> not be able to compete on a comparable basis with the United States industry. I
> do not find it surprising, therefore, that the French and the Germans are now
> considering a launcher programme. What I do find surprising and disappoint-
> ing, however, is that our own first stage rocket launcher, Blue Streak, is about
> to come to the end of its active life after only 11 firings.... If Europe is deter-
> mined to have a launcher programme, I wonder whether there is not some way
> by which the Blue Streak at least might find its way back into that programme
> and be used again. I believe that the Minister for Aerospace is right in his
> attempts to create a European space agency. Perhaps because space is such a
> new industry, it is the first that can be set up on a European basis.
>
> (McNair-Wilson, *Weekly Hansard* 1973: 831–832)

The economic and the ideological case for independent access to space for
Europe was stressed in the French National Assembly:

> The Ariane programme would guarantee the technological independence of
> [France], but nobody argues that this project pushes back the frontiers of
> science. It relies on technologies that are well known and already tried out
> by others. The cost of the launcher is currently 50% higher than the Amer-
> ican Thor-Delta and the members of the European agency have not commit-
> ted to buying it. Do I have to remind the House that the Franco-German
> telecommunication satellite 'Symphonie' will be sent into orbit by a Thor-
> Delta [American] launcher?
>
> (Boulloche, *Journal Officiel* 1974: 6376II)

The French government, cherishing the Ariane project, denied that the French
launcher was economically inferior:

> You told me that the Ariane launchings would cost more than those of Thor-
> Delta. Your data is inaccurate, because in order to launch a geostationary
> satellite of a weight comparable to that of which Ariane will have the capa-
> city, that is to say 750 kg, you would need an Altas-Centaure not a Thor-
> Delta launcher, costing more than FF80 ... so we can claim that the Ariane
> is in fact fully competitive.
>
> (d'Ornano, *Journal Officiel* 1974: 6400II)

France was able to proudly present its launcher programme as the most sophisticated in Europe. Thus Ariane was seen as the leading candidate for a European

launcher programme. This element of a European response to the ensuing space age was, in fact, a response to urging from the USA.

The post-Apollo programme: the USA calls for a 'European' response

The pressure from the USA for a 'European response' was deployed as a political argument in the discourse on the development of a European space programme.

> We felt that no consensus could be reached on a coherent European [space] programme, because of the hesitation of certain of our partners, who seem to prefer to abandon the project of developing a large European launcher, as called for by the Americans, in order to join in the Post-Apollo programme.
>
> (Charbonnel, *Journal Officiel* 1972: 4467II)

Of course, US pressure for a 'European response' put the Europeans on the spot, at a time when the European integration process was still only in its late infancy. Arguably, after rapid integration in the 1950s, the great venture actually lost ground in the 1960s, notably under the terms of the Luxembourg 'compromise' in 1966. The 1970s, when a European space policy first got under way, was a period when, to all intents and purposes, integration was otherwise marking time. Thus European space collaboration was in no danger of developing a cumbersome bureaucracy. This made a good fit with the Gaullist philosophy of 'l'Europe des patries' and also with the British preference for intergovernmental cooperation. A lean and efficient administration was the preferred mode of operation and this is what eventually led to the creation of the ESA (Charbonnel, *Journal Officiel* 1973: 5372I). European autonomy in space affairs – as can be seen in the following quotation – became, nevertheless, important and resembled to some extent a 'European response' in space affairs, as demanded by the USA:

> The Government, following my proposal and that of the Prime Minister, has decided that the essential objective of our space effort must be for Europe to enjoy autonomy in the field of space applications. In telecommunications, air and sea traffic control, in meteorology or for defence purposes, we cannot rely solely on the US or the USSR. An exclusively national effort in this field does not make sense. Our choice, therefore, must be a European one.... the European Space Agency, which will be the framework for cooperation will be an instrument of the highest standards in terms of organisation and structures.
>
> (d'Ornano, *Journal Officiel* 1974: 6364I)

The British response to a European space programme, particularly a French-led version, was lukewarm, as can be seen in the vague commitment to a European space conference and the puny financial contribution:

Mr. Bishop asked the Secretary of State for Trade and Industry what action he proposes in relation to the post-Apollo programme to preserve United Kingdom supremacy in all large structures projects and to ensure United Kingdom participation in the European Space Club, European Launcher Development Organisation and European Space Research Organisation post-Apollo programmes in the future.

Mr. Michael Heseltine: The Government have decided to take a full part, along with other European nations, in further exploratory work designed to facilitate evaluation by the European Space Conference of its response to the United States offer of participation in the post-Apollo programme. The contribution of the Government to this further work is estimated at £240,000. Much of this money will be spent with British companies. I am not able to comment as to the likely outcome of these researches until they have been completed.

(*Hansard* 1972: 29–30)

A European space conference, which assembled experts on 31 July 1973, achieved some concrete results. The participants agreed on the development of Spacelab within the US post-Apollo programme, with progress in other areas that will be discussed later (Heseltine, *Weekly Hansard* 1973: 693–694). Generally, it was accepted that the job would have to be done via European cooperation, by the British space industry (McNair-Wilson, *Weekly Hansard* 1973: 830) and by the political establishment (Onslow, *Weekly Hansard*, 1972: 177) in the UK and France alike.

These references show a competition frame in which both France and the UK see themselves in competition with the space industry in the USA. This created a 'European' response in the establishment of the ESA, although a relatively weak one compared with the supranational integration of the European Communities. The USA demanded this European response and acted one more time as a federator for Europe.

Analysis: policies and applications

In the bargaining between France and the UK, both countries tried to monopolise certain aspects of space policy.

Ariane

Closely in line with discourse theory, Ariane became the dominant launcher project, leading to the creation of the ESA. The unsuccessful predecessor programme of Europa launchers had eroded confidence in the French/European launcher programme. The UK and Italy abandoned the project in 1968–1969. Germany followed suit in 1972–1973. Only France and Belgium continued with the idea of a European launcher, but the defecting partners re-joined the so-called substitute launcher programme L III S, which became Ariane and was presented at the European space conference on 20 December 1972. Thus, in space, France chose European cooperation, but could rightly claim leadership in technological and political terms (Charbonnel, *Journal Officiel* 1973: 1299I-II, 1300I).

Proportional spending in the French budget shows the importance of individual space projects. Ariane was by far the biggest spend in the French space budget (Charbonnel, *Journal Officiel* 1973: 5372I). It was supposed to deliver independent access to space for Europe and, of course, France (d'Ornano, *Journal Officiel* 1974: 7101I). The innovative character of the development of a heavy launcher was, however, doubted, because the technology had been around for a long time (Boulloche, *Journal Officiel* 1974: 6376II). France could not finance Ariane alone. Therefore European cooperation on the project was vital (d'Ornano, *Journal Officiel* 1974: 6364I).

The British Government agreed and was even prepared to contribute funding to the French-led project, but only in exchange for French funding for the British-led Maritime Orbital Test Satellites (Marots) (Heseltine, *Weekly Hansard* 1973: 49). This established, as in other areas, the precedent of *juste retour*, under which a country receives roughly the same amount of money given to the ESA back in contracts for its own space industry. Spending issues were consequently the major concern (Kaufmann, *Weekly Hansard* 1975: 563–564). Thus there were strong national interests at play, but also a potential for a truly European space industry, not least for launchers (McNair-Wilson, *Weekly Hansard* 1973: 831–832).

Satellites

Satellites such as Marots were the most important, because they were potentially the most lucrative of the space applications. Marots became one of the founding projects of the ESA (Heseltine, *Weekly Hansard* 1973: 693–694) and the founding nations, including France, contributed to the funding of this satellite. This was the trade-off for British support for the French Ariane launcher and the financial contributions of each nation for the other's project were reciprocal. France paid into the Marots satellite the same amount as Britain paid into the Ariane programme (Heseltine, *Weekly Hansard* 1973: 49). However, Britain clearly remained the main advocate for European investment in space (Onslow, *Weekly Hansard* 1973: 842).

The European Space Research Organisation (ESRO) had developed a number of satellites, e.g. Meteosat for meteorology, a European telecommunications satellite and a navigation satellite for air traffic (Charbonnel, *Journal Officiel* 1973: 5372I). The problem in the period before the ESA was that most European satellites, notably the Franco-German telecommunications satellite Symphony, were sent into orbit by either US or Soviet launchers. This might not seem such a big problem for civil applications, but once power considerations come into play, antagonistic rationales vis-à-vis the other major powers were invoked, which ultimately led to the French claim for independent access to space for Europe (Boulloche, *Journal Officiel* 1974: 6376II; d'Ornano, *Journal Officiel* 1974: 7101I).

This section has shown the discourse relating to the pre-eminent technologies, installations or innovations that both France and Britain 'contested'. In most areas, the French solutions prevailed, but participation in the space discourse meant that the UK remained on board, sharing the final outcome to a considerable extent, as seen with the political commitment for the ESA.

Analysis: polity organisations

A number of organisations were predecessors to the ESA. They helped formulate a European approach, but fell by the wayside in the process. It is nevertheless important to understand the rationale for their creation to attain a better understanding of the process that led to the creation of the ESA.

European Space Research Organisation

Established in 1964, the ESRO was one of the first attempts to bring the diverse and often overlapping European space programmes under a shared umbrella. Such national space programmes were finally grouped under three main branches of satellite development: telecommunications, air traffic control and meteorology. France was the lead nation in this organisation, with a clear agenda of Europeanisation of space assets across Europe (Charbonnel, *Journal Officiel* 1972: 1167II). The UK and its space industry followed this lead:

> However, those I have spoken to in our industry generally give the concept of the ESA a welcome, even if be a qualified one. They see it as absorbing the work being done by ESRO ... but I believe they wonder to what extent the ESA will be given powers to have a firm direction over Europe's space activities and thus to be cost effective. Certainly they see that nothing is to be lost by co-operating with the agency as regards informing it of national projects. They also think that the co-ordination of programmes within Europe is to be welcomed.
>
> (McNair-Wilson, *Weekly Hansard* 1973: 830)

The hope was that, in the ESA, duplication of space research could be avoided across Europe without overbearing bureaucracy (Onslow, *Weekly Hansard* 1973: 840). Technological development seemed to warrant hope for growth in the space industry because, particularly in the satellite programmes under ESRO, there was maturation in the early 1970s seen as a shift from using satellites for scientific observation to using them for commercial services in communications and other applications. This change offered strong indications that the industrial significance of space activities would increase further (Onslow, *Weekly Hansard* 1973: 836). In commercialisation and in the avoidance of duplication, we can also discern the hegemonisation of nodal points, i.e. the filling of such empty signifiers with space contents.

European Launcher Development Organisation

The European Launcher Development Organisation (ELDO), founded in 1962 and charged with the development of a European launcher, was also a forerunner of the ESA. Its history was marred by the failures of the Europa rockets, based on the Cecles-ELDO engines (Charbonnel, *Journal Officiel* 1972: 4467II). Overspending and successive failures of these launchers led to the demise of ELDO in 1973 (McNair-Wilson, *Weekly Hansard* 1973: 831–832; Cousté, *Journal Officiel* 1972: 4468I). For the short period until the advent of the ESA in 1975, ESRO took over its tasks, administration and facilities.

During this early period of space endeavours, the nationalist impulse can be felt fairly strongly. A European spirit of collaboration in space technologies still had to develop:

> Mr. Bishop asked the Secretary of State for Trade and Industry what action he proposes in relation to the post-Apollo programme to preserve United Kingdom supremacy in all large structures projects and to ensure United Kingdom participation in the European Space Club, European Launcher Development Organisation and European Space Research Organisation post-Apollo programmes in the future.
>
> (*Hansard* 1972: 29–30)

European collaboration aspects were, however, developing. The forum where such ideas found the right environment was the European space conference, held on 20 December 1972, followed by a further such meeting on 31 July 1973, clear evidence of the urgency and perhaps the real importance of space affairs at the time (Heath, *Weekly Hansard* 1972: 1399). ELDO was by that time already close to burn out, with Italy withdrawing as early as 1969 and the main contributors, France, Germany and the UK, squabbling about its future (Charbonnel, *Journal Officiel* 1973: 1299I-II, 1300I). The European space conference brought the presentation of a new launcher programme, under several names: the L III S, substitute launcher for the European programme and, finally, the Ariane launcher. The conference yielded an agreement to establish a European space agency that would encompass the work of both ELDO and ESRO.

> After the uncertainties and the hesitations during 1972, the successive failures of the last launches of the Europa II rocket; and after difficult and painful decisions in the abandoning of the Europa III programme and then the Europa II; after long and difficult negotiations, we were able, it is no exaggeration, to rescue the European space endeavour. We were able at last to give Europe a comprehensive space programme, coherent and balanced. 'Europe' I say, quite advisedly, because in this field our policy is uncompromisingly open to international cooperation and naturally we give preference to cooperation with our European partners.
>
> (Charbonnel, *Journal Officiel* 1973: 5371II-5372I)

This led the way to the establishment of the ESA, which can be seen as the outcome of the hegemonisation process working in the field of European space organisation between European nations.

European Space Agency: a civilian organisation

The European integration process of the early 1970s was no longer the idealistic 'brave new world' onward march of the 1950s, when Jean Monnet cried out for the creation of a United States of Europe (Monnet 1976). The structure of the ESA reflects the changed spirit bought about in France by the return of a passionately nationalist de Gaulle, the Fifth Republic and the new leader's idea of 'l'Europe des patries'. The ESA was designed as an intergovernmental institution, bereft of political aspirations of any kind. It was meant to be an effective executive agency of the European space programmes, avoiding duplication between them (Charbonnel, *Journal Officiel* 1973: 5372I; d'Ornano, *Journal Officiel* 1974: 6364I; Heseltine, *Weekly Hansard* 1973: 693–694). The staffing was to be kept very small to avoid overspending, but also to prevent the development of an authority with a mind of its own, independent of the Member States:

> Turning now to the European Space Agency concept, the point made by the hon. Gentleman that one could have a proliferation of bureaucracy is right. Certainly, we do not want to set up an agency which would grow – I think it is called Parkinson's Law – without there necessarily being anything for it to do.
>
> (McNair-Wilson, *Weekly Hansard* 1973: 829)

Importantly, the ESA was also designed as an exclusively civilian organisation (McNair-Wilson, *Weekly Hansard* 1973: 831). The European space conference of 31 July 1973 agreed to the establishment of the ESA and it agreed action for the future harmonisation of national and European space programmes. At this point, the ESA was meant to start work on 1 April 1974 (Heseltine, *Weekly Hansard* 1973: 693–694). In fact, the Member States only signed the Convention in May 1975, with proper ratification of the Treaty in October 1980. This long period of gestation can be explained by the persistence of strong nationalist positions, strongly argued by a number of Member States (Dalyell, *Weekly Hansard* 1973: 825). This also required a good deal of give-and-take within the Member States to enable the governments to win majorities endorsing the Treaty. Despite the fact that the ESA was called the *European* space agency, this section has shown that national interests were never far from the minds of the players involved.

Competing frames

Analysis of primary sources from the period in France and the UK reveal that both countries agreed to the foundation of the ESA. That does not mean, however, that national interests were eclipsed. They persisted, thrived and became bargaining frames in the overall discourse on the development of a European space policy. This sounds almost a little too integrationist, because it can well be argued that the sharing of the discourse did not take place everywhere. For the UK, the main concern remained the question of costs; for France, the main concern remained the possession of independent facilities, notably independent of the USA. However, both countries found a common denominator in the development of a viable European space industry.

Britain's budget frame

A good insight into British thinking at the time is given by the following quotation, which begins by making the idealistic point that there should be a European space programme, but then turns to real interests in the British space industry and the British funding that would have to be found for the ESA:

> What I think my hon. Friend the Minister [Member for Tavistock (Mr. Michael Heseltine)] is rightly trying to do is to make Western European countries think in European rather than in national terms. As we are such a new member of the EEC, it is heartening to see the younger Ministers pushing forward with a concept of European agencies rather than struggling on with national concepts that do not, and cannot measure up to those of our competitors, in particular our North American competitors.... It is not unreasonable to suppose that ... we could leapfrog from a national space agency and go straight to a European Space Agency ... there is some way to go before a European Space Agency gets off the ground ... should [we] not consider again the Select Committee's recommendation that a national space agency bears serious consideration.... In other words, Britain may have a number of separate projects in hand, but she has no overall space programme as such. As we are spending more than £30 million a year on space, one may wonder whether we are getting value for money.
>
> (McNair-Wilson, *Weekly Hansard* 1973: 828–829)

Budget concerns are the main frame in the British debate on a European space programme. However, it was also a moment of creation. Serious questions about the philosophy of the new European space agency were being asked:

> Equally, this is a moment to consider what we think the European Space Agency should be about and what should be its terms of reference. Do we accept the concept that all European civil space projects should be handled or coordinated by a single agency; and do we think that member countries should commit their entire civil space budgets to it? That seems a fairly tall order. If, perhaps, we question whether that is how it should operate, do we think that nations should contribute on a GNP-related block sum basis, or do we think that they should make pro rata contributions only to those programmes in which they are involved?' I gather that the general European

view is that individual nations should receive back at least 70 per cent of any contributions they make in contracts placed by the agency.

(McNair-Wilson, *Weekly Hansard* 1973: 830)

Finally, the argument turns again towards the ESA budget and how it should be organised. The key organisational question in this respect was whether the budget should be made up of fixed contributions based on the GDP or whether the countries should pay only for the programmes they were interested in – often called an 'à la carte' approach (Onslow, *Weekly Hansard* 1973: 840). This question was settled in the more general question on the ESA budget and the British contribution to it:

> Mr. Michael McNair-Wilson asked the Secretary of State for Industry on what the European Space Agency's budget of £185 million is being spent; on what basis Great Britain's contribution has been decided; which countries are paying as much or more; and what work is being carried out in the British aerospace industry as a result of our membership of the ESA. Mr. Kaufmann: About 85 per cent of the budget of the European Space Agency is devoted to applications programmes covering development and proving of communication satellite system technology, meteorological satellite, a spacecraft launching vehicle, and a space laboratory to be carried in the United States space shuttle. The remainder is being spent on scientific research in space; the United Kingdom contribution to this is the responsibility of the Science Research Council. Contributions to the majority of the ESA programmes are on a gross national product basis, although the three most recently adopted applications programmes are being funded according to the national interests of the countries participating in each specific programme.
>
> (*Weekly Hansard* 1975: 563–564)

Thus Britain achieved budgetary flexibility in the ESA, a funding of British space projects roughly equal to what Britain paid in, and therefore a financial contribution entailing little in the way of budgetary strain.

France's independence frame

The main French concern was independence, mainly in political terms, but also in terms of developing a space industry that would be competitive in cutting-edge technologies:

> France has thus deliberately chosen – I can confirm this to the House ... the road to European independence in this field [space]. We fully realise that this is a narrow and difficult road, but it is the only one that has a future and corresponds to the dignity of our continent.
>
> (Charbonnel, *Journal Officiel* 1972: 4927II, see also, 4467II)

The development of a European launcher programme, first in the Europa rockets, later supplemented by Ariane, fitted well with the argument of European independence in that such launchers would give Europe independent access to space, vital for its political freedom to pursue its space aspirations unhindered (Cousté, *Journal Officiel* 1972: 4468I; Charbonnel, *Journal Officiel* 1973: 1299I–II, 1300I; see also 5371II–5372I). The problem was that 'European' (in the sense of the European Communities) was still a very risky term to use in space affairs, because the Treaties of Rome provided no specific powers for the European institutions in this field and thus it remained up to the will – or whim – of the Member States whether to cooperate (Cousté (JO), 1973: 1300I). The ESA was set up on an intergovernmental basis and this relatively unstable arrangement ultimately turned out to be the organisation's Achilles heel. We could see this as a shortcoming resulting from the political will, not least of France, at the time. France had, however, achieved its main objective of independence in the space sector (d'Ornano, *Journal Officiel* 1974: 6364I, 7101I).

The European frame: a European space industry

This can be seen in the existence of a European space industry, still based on national interest, not least in national space companies. They are internationally competitive, as was the original objective (d'Ornano, *Journal Officiel* 1974: 6400II), because of European cooperation, which goes as far as the actual naming of European companies, such as Arianespace, founded in 1980. The French aspiration to independence was thus just as much an assertion of political independence as a claim to technological independence, which would be embodied in a European space industry producing satellites, launchers and technology for research, such as the Spacelab (d'Ornano, *Journal Officiel* 1974: 6364I). British commitment to involvement in this endeavour was also forthcoming (Heseltine (Han) 1972: 29–30):

> We are in the EEC. If we have to accept this kind of marriage, it is not good enough for the United Kingdom not to take a lead in these important matters.... Many people will not be able to see the relevance of putting men on the moon. But in this matter we are not concerned with such ambitious enterprises. However, there is an enormous spin-off of high technology from post-Apollo projects and the Government and the industry must spell out the benefits of space participation in terms of medicine, biology, counter-pollution measures, meteorology, weather forecasting, telecommunications, crop control and many other aspects which will help not only the developed, but the under-developed, countries. This is a job for the Government and industry.
>
> (Bishop, *Weekly Hansard* 1973: 834)

The progress European space efforts had made since the inception of the early space organisations ESRO and ELDO was substantial. The shift from theory to applications made the commercialisation of space services possible and held out the prospect of a growing space industry in Europe (Onslow, *Weekly Hansard* 1973: 836). Both France and the UK wanted their fair share of this growing sector. In the principle of *juste retour*, we can discern the safeguard that France and the UK wrote into the ESA Convention that their taxpayers' money would be returned in due course through the development of their home-grown space industries.

Conclusion

It can be concluded that European space policies moved initially from disorganised national space policies to ad hoc space collaboration between nations under arrangements such as those of ESRO and ELDO. This was quickly recognised as an unsatisfactory development that did not yield the desired results. Therefore ten European states[1] eventually agreed to the creation of the ESA in 1975 to cooperate in the field of space. This membership has now expanded to 18, plus Canada.[2] Space policies became more coordinated as a result, but the organisation remains intergovernmental in accordance with the will of its members. There are some integration aspects, such as the GDP contributions to the ESA budget rather than the exclusively 'à la carte' financial contribution originally mooted. However, the hopes that the ESA might develop into another supranational institution were soon dashed. It became clear relatively quickly that space had strong connections with, and indeed had its origins in, the aerospace and defence industries. However, there was, of course, also a definite degree of real innovation, as can be seen in the agreement between France and the UK to foster a European space industry. This helped to establish a more natural collaboration between the respective national space industries than would have been possible in the defence sector, for example, where secrecy and vested interests prevail to this day.

This chapter has thus sketched out the potential of the space sector as a component of the greater European integration movement. This has led to the foundation of the ESA because the nation states could not afford viable national space policies of their own, thus being impelled towards a European solution, albeit not a political one, much less a supranational one. The ESA was deliberately left apolitical and intergovernmental, although this might have to change in the future if space policy assumes a more prominent role beyond its technical aspects. This would require political legitimacy – a rationale that can be seen in the current political debate on the relationship between the ESA and the EU (see von der Dunk 2003; Hobe 2004; Verheugen 2005; Gaubert and Lebeau 2009; Peter and Stoffl 2009; Hörber 2009).

Notes

1 Belgium, Denmark, France, Germany, Italy, the Netherlands, Spain, Sweden, Switzerland and the UK were the founding members of the ESA.

2 Austria, Finland, Greece, Ireland, Luxembourg, Norway, Portugal and the Czech Republic joined later. Since 1 January 1979, Canada has the special status of the ESA 'cooperating state'.

Bibliography

Primary sources

More detail about the exact location of primary sources can be provided by the author, e.g. exact dates of quotations or volume numbers of the *Journal Officiel*, *Hansard* or *Weekly Hansard*.

Hansard 1950–1959, Vols. 472–628.
Hansard 1972, Vol. 835.
Journal Officiel (abbreviated in the sources as JO), Assemblée Nationale, First to Third legislature, Paris, 1946–1958.
Journal Officiel (abbreviated in the sources as JO), Fourth legislature from 11 July 1968 to 1 April 1973, Assemblée Nationale, Paris, 1972–1973.
Journal Officiel (abbreviated in the sources as JO), Fifth legislature from 2 avril 1973 to 2 April 1978, Assemblée Nationale, Paris, 1973–1975.
Weekly Hansard 1971–1975, Issue No. 843–1019.

Secondary sources

Dunk, F. von der (2003) 'Towards one captain on the European spaceship – why the EU should join ESA'. *Space Policy*, 19(2): 83–96.
Gaubert, A. and Lebeau, L. (2009) 'Reforming European space governance'. *Space Policy*, 25(1): 37–44.
Hall, P. (1992) 'The movement from Keynesianism to monetarism: institutional analysis and British economic policy in the 1970s'. In: S. Steinmo, K. Thelen and F. Longstreth (eds), *Structuring Politics.* Cambridge and New York: Cambridge University Press, 90–113.
Hall, P. (1993) 'Policy paradigm, social learning and the state'. *Comparative Politics*, 25(3): 275–296.
Hall, P. (1997) 'The role of interests, institutions and ideas in the comparative political economy of the industrialized nations'. In: M. Lichbach and A. Zuckerman (eds), *Comparative Politics.* Cambridge: Cambridge University Press.
Hall, P. (1998) 'The political economy of European an era of interdependence'. In: H. Kitschelt *et al.* (eds), *Change and Continuity in Contemporary Capitalism.* Cambridge and New York: Cambridge University Press.
Hobe, S. (2004) 'Prospects for a European space administration'. *Space Policy*, 20(1): 25–29.
Hörber, T. (2009) 'The European Space Agency (ESA) and the European Union (EU) – the next step on the road to the stars'. *Journal of Contemporary European Research*, 5(3): 405–413.
Jobert, B. and Muller, P. (1987) *L'etat en action.* Paris: Presses Universitaires de France.
Kohler-Koch, B. (2000) 'Framing: the bottleneck of constructing legitimate institutions'. *Journal of European Public Policy*, 7(4): 513–531.
Laclau, E. and Mouffe, C. (1985) *Hegemony and Socialist Strategy.* London: Verso.

Mörth, U. (2000) 'Competing frames in the European Commission – the case of the defence industry and equipment issue'. *Journal of European Public Policy*, 7(2): 173–189.

Monnet, J. (1976) *Mémoires*. Fayard: Paris.

Peter, N. and Stoffl, K. (2009) 'Global space exploration 2025: Europe's perspectives for partnerships'. *Space Policy*, 25(1): 29–36.

Rein, M. and Schön, D. (1991) 'Frame-reflective policy discourse'. In: P. Wagner (ed.), *Social Sciences and Modern States: National Experiences and Theoretical Crossroads*. Cambridge: Cambridge University Press, 262–289.

Sabatier, P. (1993) 'Policy change over a decade or more'. In: P. Sabatier and H. Jenkins-Smith (eds), *Policy Change and Learning*. Boulder, CO: Westview Press, 13–39.

Sabatier, P. (1998) 'The advocacy coalition framework: revisions and relevance for Europe'. *Journal of European Public Policy*, 5(1): 98–130.

Surel, Y. (2000) 'The role of cognitive and normative frames in policy-making'. *Journal of European Public Policy*, 7(4): 495–512.

Verheugen, G. (2005) 'Europe's space plans and opportunities for cooperation'. *Space Policy*, 21(2): 93–95.

2 Framing as a tool for analysing European space policy

Paul Stephenson

Introduction

The first two Galileo satellites were launched by a Russian Soyuz rocket from a base in French Guiana on 21 October 2011. This marked a crucial and long overdue step in the EU's multi-billion euro investment in its own version of the US global positioning system. Galileo is expected to bring significant returns to the EU economy in the form of new businesses that can exploit precise space-borne timing and location data. According to a press release published that day on the European Commission's own website, the satellite navigation sector had already become 'very important' for the EU economy (contributing roughly 7 per cent of EU GDP in 2009) as well as for the 'well-being of its citizens' (see Europa website).

It took three hours and forty-nine minutes for the satellite pair to reach their correct orbit 23,222 km above the earth. The BBC reported that European Communities Vice-President and Commissioner for Industry and Entrepreneurship, Antonio Tajani, said: 'Galileo is at the heart of our new industrial policy.... We must commit very strongly to Galileo. We need this; this is not entertainment. This is necessary for the competitiveness of our European Union in the world.' He then announced the industrial competition to procure six to eight more satellites over and above the 18 already contracted (Amos, 2011).

The long overdue deployment phase of Galileo was the culmination of a lengthy process of framing and reframing the issue of satellite navigation and the question of whether Europe should pursue a collective space policy. A constellation of satellites is now planned for completion by 2019. The European Commission is currently promoting the message, based on independent studies, that Galileo will deliver around €90 billion to the EU economy in the first 20 years of operation 'in the form of direct revenues for the space, receivers and applications industries' and in 'indirect revenues for society (more effective transport systems, more effective rescue operations, etc.)' (see Europa website). Such an estimate may appear wholly ambitious for some in the space industry. Nonetheless, this 'economy' frame, which seeks to convey the benefits to 'users', 'receivers' and 'industries', is just the latest way of presenting the issue of satellite navigation as the prospect of fully exploiting the system draws closer.

This chapter examines the frames constructed and articulated by the Commission alone, to see how the discourse of the (different parts of the) agenda-setter have changed over time and whether or not there has been conflict and controversy. Such an analysis should illustrate the dynamics of intra-institutional relationships and identify the role of the leadership in asserting convincing frames. Because proposals initiated by the Commission often result in decisions and Resolutions by the European Parliament and Council, the frames dominating the EU's policy discourse are largely constructed by the Commission. It does not do this in isolation, however, but as the result of communication and liaison with experts in the European Space Agency, as well as private investors.

The second section makes a case for studying frames within policy discourse. The third section looks at how frames might be relevant to the agenda-setting activity of the European Commission when formulating proposals for new legislation, as well as at the different dimensions of frames worthy of examination. The fourth section provides an in-depth case analysis of frames within the policy discourse on Galileo at three points in time: 1992, 1996 and 2000 (Commission 1992, 1996, 2000). The fifth part, the conclusion, sums up the main observations regarding the evolution of frames, while also reflecting on how this example contributes to the study of frames when shaping policy.

Policy frames and institutional discourse

Frames and policy-making

What exactly is Galileo about? What is it a policy solution for? Or, from a different angle, what is the issue for the EU? What is the 'problem' to be solved? Schön and Rein assert that interpreting social reality implies the social construction of policy problems (Schön and Rein 1994: 31; Rein and Schön 1993: 145–166). Policy-makers construct policy problems as well as offering policy solutions by framing issues. They do this by taking into account the assumptions held by participants in forums where policy discourse 'circulates' and is 'constructed', as well as acknowledging hierarchies of power within the institutional architecture of the policy-making system – action is embedded in a structural context that limits the scope for action. As such, policy-makers act as designers, engaging in dialogue with experts and interests, and play an active role in the construction of the images and venues used to communicate issues. An issue is thus the subject *of* policy discourse, but is also subject *to* policy discourse, particularly where there are framing conflicts and controversy (Schön and Rein 1994).

The work of Dryzek (1993) makes a case for policy analysis as the examination of argument, whereas Fischer (2003) sees it as a discursive and deliberative practice. Indeed, the seminal work on framing by Goffman (1986) analysed the organisational experience of policy-making as a succession of cultivated and projected frames. Policy-making is itself a process of political contention where multiple interests may hold conflicting interests and perspectives. Even in the same institution there may be conflicts between units, and thus competition to frame the issue most convincingly. Frames are valuable because they alter perceptions, beliefs and appreciation, determining how people understand an issue and what for them counts as 'fact'. As such, framers establish which arguments are considered relevant and compelling (Goffman 1986). There may be rivalry both between and inside institutions to assert frames to control the process, hence we can envisage institutional struggles to name and frame a policy 'solution' – that is, the solutions to what exactly? Whose policy remit will the solution affect? As a result, we might expect to see symbolic (discursive) contests over the social and political meaning of an issue, because meaning ultimately determines the policy action through decision-making (or no decision), with political decisions legitimising choices over the use of finite budgetary resources.

Frames held by actors determine what they see as being in their interest – actors 'sponsor' frames. These frames are not free-floating, but are grounded in institutions, which have their own interests to advance. In a world of frames there is no objectivity or falsifiable frame, only subjectivity, where those seeking to sell their frame must make sense of complex, information-rich environments. This requires selecting and sorting data, then organising it in such a way that it is palatable, digestible and filling (Goffman 1986).

Frames and institutions in the EU

Discourse theory and framing have been addressed by a range of European integration and public policy scholars, such as Kohler-Koch (2000), Surel (2000), Weaver (2000), Torfing (2005) and Daviter (2007). It has also been applied to a range of EU policy areas, including steel (Dudley and Richardson 1999), labour and welfare (Dostal 2004) and the Common Agricultural Policy (Lynggaard 2007), as well as in areas with a clear link to space policy, such as defence (Mörth 2000) and common foreign security policy (Smith 2003).

The European Commission, Council and Parliament each has their own discourses, their competencies determined by the EU Treaties. The Commission is to all extents and purposes the agenda-setter, although the European Council (summits of Member State leaders), was also recently given formal status at Lisbon to set the agenda. The Commission thus makes proposals (often at the request of others) after consulting with experts and interest groups. What is important to acknowledge, however, is that the Commission is itself not an institutional monolith, but rather a differentiated body consisting of many departments (Directorate-Generals being the link with ministerial departments at the supranational level). As such, how an issue is framed will ultimately depend on which part of the Commission can claim that its remit extends to that issue – for example, whether it is a trade or transport, or an internal market issue. In some instances, several Directorate-Generals will feel they should be dealing with the particular issue and, in so doing, they may construct their own frames so that others perceive the issue as pertaining to their own policy domain and, thus, 'jurisdiction'. Thereafter, frames may be modified over time to secure greater salience for the issue or to try to push it higher up the policy agenda.

Mörth (2000) has explored the competing frames operating within the Commission for the example of the defence industry. Her work on the Commission as a 'multi-organisation' found that different frames compete across policy areas – and hence across Directorate-Generals – often revealing conflicts of interest. She found that frames were not static, but changed over time, which led to reframing. Her later work on the efficiency and accountability aspects of public–private partnerships in Galileo concluded that the policy had many goals, none of which was easy to combine, implying the existence – and need for – multiple frames to operate at any time to appeal to a wide range of stakeholders (Mörth 2007). Thus frames may compete within an institution, meaning that multiple, often conflicting, frames circulate in the policy-making environment, each informed by one or more sets of experts and interest groups. In the European Commission some higher profile Directorate-Generals may inherently be at an advantage when it comes to 'housing' issue frames, merely because their area of work is deemed to have greater political stakes. For example, Directorate-General Competition or Internal Market may have a higher profile than Directorate-General Health and Social Affairs or Fisheries. If an issue is high up the political agenda, all parts of an institution may claim ownership or assert that their 'solution' is best.

Frames and the EU's institutional architecture of space policy

Any analysis of frames requires an understanding of the institutional architecture of the 'framers' in order to map institutional discourses and how they emerge. In the EU system, we can imagine variations in the way that satellite navigation is framed between the different Council working groups, the Parliament committee and the Commission Directorate-Generals (Transport, External Relations, Industry and Entrepreneurship). Moreover, the European Space Agency (established 1975) and the European Defence Agency (established 2004) have their own discourses, as do bodies such as the Space Councils, which bring together the Council of Ministers and the European Space Agency, and the Trans-European Networks Executive Agency (established 2006), which is responsible for implementing the EU's priority infrastructure projects. Bodies in charge of financing satellite navigation also have distinctive frames, such as the European Investment Bank (established 1958), which provides loans, and the European Global Navigation Satellite System Supervisory Authority, a new Community Agency that officially took over responsibility from the former Galileo Joint Undertaking on 1 January 2007.

Framework for analysis

Analysing Commission storytelling

Analysing how issues are discussed and projected, and therefore heard, read and perceived, means tracing the evolution of frames over time. Any understanding of the way in which Galileo is talked about and legitimised today requires working backwards to see how the story has evolved. There are limits within the scope of this chapter as to how closely we can put official documents under the microscope in terms of qualitative and quantitative analysis. Approaches to systematic discourse analysis range from the examination of syntax and the use of metaphor to actually counting the frequency with which words are used. Van Dijk (1977) distinguishes between the study of language (a domain including lexicons, semantics, grammar, syntax and macro/discursive structures), which he defines as the study of the parts people use to construct a text, and the study of discourse (a domain of 'semantic superstructures' or 'schematic forms', such as narratives, myths, arguments or scientific reports), as cited in Fischer (1997).

As such, this analysis will seek to identify the evolution of the main frames over time by considering 29 official documents, all authored by the Commission and pertaining to space, satellite navigation and, specifically, to Galileo, while honing in more closely on five individual texts. These range from the first general Commission Communication from 1988 *The European Community and Space: a Coherent Approach* (Commission 1988) to the recent 2011 communications *A Budget for Europe 2020* (Commission 2011a) and *A Growth Package for Integrated European Infrastructures* (Commission 2011b). Communications are published 'to' the Council and European Parliament and include a number of proposals, action plans and progress reports. While other publication types mentioning space policy are also accessible, such as Commission staff working documents and White Papers, to be consistent in the analysis of structural features and frames it is necessary to look at the same document type, authored with specific institutions (recipient readers) in mind and to acknowledge that the documents were for external audiences. By identifying the early frames within the institutional discourse on Europe and space, this chapter analyses how these frames have evolved over time in line with policy content and why this may have been so.

Structural aspects of policy communications

It is important to acknowledge certain structural features of policy or scientific reports in general, which may be common to all such documents regardless of policy area, pertaining to the way information is structured, codified, sequenced, laid out and signposted. For an agenda-setter, whose communications are often proposals advocating a course of action, we can assume the use of persuasive and convincing language that may often use emotive and symbolic language, including analogy and metaphor.

A text's surface structure contains codes, hypercodes and signature elements, as referred to by Fischer (1997, citing conference papers by Donati and Triandafyllidou). Codes are marked lexicons and text segments that signify a deeper meaning beyond the basic lexicon of the word or words in a phrase in isolation. Hypercodes are collections of words that give meaning to codes and signature elements alert the reader/listener to interpret a text in terms of a particular frame. The surface structure also contains discursive structure, which organises the meaning of passages of texts. These are terms originally proposed by van Dijk (1977), according to Fischer (1997). These include rhetorical devices, use of mood, verb tense and the sense of voice that may be unique to the Commission as an institutional actor. As well as surface structure, there is deep structure, which includes narrative structures, or storylines and plots, and ideologies.

Frame dimensions

How can we deconstruct the process by which an issue is framed? This can be achieved by examining how problems and solutions are articulated with respect to an issue, while also acknowledging the overriding values and beliefs of the policy-making system. According to Radulova (2011: 41–43), the close analysis of documents can reveal the existence of one or more frames, each presenting the issue in a different light, while leaving other possible representations in the shadow. Each frame thus defines a problem separately and differently, structuring information and impressions to provide a succinct, easily digestible message. Frames are thus selective ways of shaping perceptions and understanding – often those of people who need to be convinced and who possess competence to decide in favour of the issue, or to support it financially. Frames project a 'diagnosis' (imply there is a problem) and a 'prognosis' (suggest a solution). As we know from Kingdon (2003), solutions may often exist first and then go looking for problems. For these two frame types, we might also consider motivational framing, or rallying the troops behind the cause, as referred to by Snow and Benford (1988: 199–202).

Frame analysis is useful for understanding the evolution of frames or, perhaps, the journeys that they take. By tracing this journey, we can glean insights into what happened and why. Radulova (2011) puts forward a four-dimensional framework for examining the inter-relations between policy frames: a normative, constitutive, cognitive and policy dimension. The first dimension – normative – refers to the basic judgements made upon, and values attached to, social (political) reality. This implies morally determining what is desirable and taking a principled, normative standpoint. This relates to what Snow and Benford (1988) call frame resonance, whereby movements must appeal to the existing values and beliefs, or the frames will not strike a chord with the target population. The second constitutive dimension refers to the process of identifying problems – which implies agency – and relates to institutional beliefs and political ideologies. The third agent (institution) then needs to communicate the problem through a narrative that conveys the notion of causality, making others understand that there is a cause–effect relationship at play, hence the cognitive dimension. The fourth dimension, policy, outlines a course of action to address the (unfortunate) situation.

For the remainder of this chapter, the first part of the analysis will seek to draw out some main structural features of the Commission communications with regard to their codes and signature elements, as well as their deep structure – that is, the storylines that persist. Thereafter, assuming that sufficient analysis of written documents can successfully distinguish and separate out the frame dimensions, the second part of the case analysis will seek to identify different competing or coexisting frames over time.

Case analysis

Structural elements

Surface structure: codes and signature elements

The first communications on a navigation and satellite system from 1998 and on Galileo from 1999 were both extensive, at over 50 pages each, including four-page executive summaries (Commission 1998, 1999). They contain formal cover pages with a distinctive pink left-hand margin and then a contents page featuring the official code number for the document. They are both lengthy texts discussing the issue in the first half, with the second half consisting of a series of annexes featuring tables of action plans alluding to organisational responsibilities, indicating deadlines and elaborating financial cost projections.

Surface structure: discursive and organising elements

The first communication 'proposes a strategy for the EU for ensuring a European dimension to the GNSS' (Commission 1998), while the second 'sets out a strategy to secure a full role for Europe in the next generation of GNSS' (Commission 1999). The communications open as if they are invitations to a course of action: 'The EU institutions and Member States are invited to endorse the strategy proposed' and thereafter 'The Community institutions are invited to [take up the Commission's recommendations]'. The executive summaries are organised into sections pertaining to the overall strategy, scope for international cooperation, what system to choose, finance, implementation, organisational issues and recommendations. The main body of the documents share common features in that they have early sections concerning 'The challenge facing Europe' (subpart 'the issues at stake'), followed by 'The EU role' and 'Strategy for the EU' – that is, if Europe must act, then the EU will do so on its behalf. Both documents also explicitly acknowledge 'the other' by examining Europe vis-à-vis both the USA and the Russian Federation and discussing the potential for partnership and joint systems development. Timing, finance (cost-effectiveness) and technical features are all considered, as well as the organisational framework. Both documents end by discussing implementation and the way forward.

The discursive tone is formal, with 'the Commission' referring to itself in the third person. Other such rhetorical devices seek to maintain a neutral tone, so that the content appears objective and factual when, in fact, it may be subjective opinion: 'The issue is not therefore...'; 'As the communication makes clear...'; 'It is recognised that...'; and 'The principal advantages of this approach would be...'. Nonetheless, the Commission does assert what it thinks and claims ownership for ideas: 'The Commission identified/recognised/was requested to explore/will soon...'. Bullet points are used, often to announce in a succinct manner what the Commission has already identified and then what it proposes – for example, a three-point financing strategy, three basic organisational levels, three broad options, or three further questions needing to be answered.

Other key discursive tones featuring throughout are: (1) collective problem – 'The EU is faced with...'; (2) large scale – 'a formidable challenge'; (3) vital importance – 'systems which play a crucial role'; (4) positive future – 'a major opportunity'; (5) obstacles to overcome – 'serious problems of sovereignty and security'; (6) timeliness – 'an urgent decision is needed'; (7) achievements so far – 'work by the Commission over the last year has focussed on...'; and (8) role definition – 'Europe must be...'.

The communications often give the EU an anthropomorphic identity, employing a body metaphor: the EU must continue to 'embrace'; 'the EU is faced with...'; it should 'gain an initial foothold'; 'there is a need to ensure European users are not hostage to...'; '[the EU] with a clear view on (seeing)...' and 'earmark the necessary public funding'. A second marker that runs throughout is the race metaphor: 'The EU is now in a position...'; [the US] 'has a head start'; 'the Commission must press ahead'; and 'Europe's capacity to compete'. A third metaphor – though arguably featuring inherently in many policy domains – is the game metaphor: 'issues at stake'; 'having a major role'; 'other key players'; and 'risks and opportunities'.

The language is generally neutral in tone, although occasionally colloquial phrases creep in: 'the shape of Europe's "best buy" is now relatively clear'; 'or do we decide to "go it alone"?' What is striking, however, is the multiplicity of positive verbs advocating types of action – for example, the EU/Member States should 'agree', 'confirm', 'decide', 'demonstrate', 'develop', 'endorse', 'ensure', 'identify', 'intensify' and 'recognise'. Moreover, adjectives reinforce the status of the desired outcomes – 'a robust Galileo'; 'firm political decisions'; 'a key priority'; 'considerable economies'; 'lucrative market'; 'clear/major opportunities'; 'appropriate financing framework'; 'maximum private sector involvement'; 'dense schedule of meetings'; 'unique project/character' – while also amplifying the negative implications of a failure to act – 'excessive cost or risk'; 'Europe entirely dependent'; 'serious problems'; Europe risks being 'constrained', 'undermined', 'too late', so its Member States must be 'flexible', 'confident' and 'unequivocal'.

Deep structure: narratives/storylines

There are five central narratives running through the communications, which can be understood as sequential even if they do not appear explicitly so. The first concerns the problem – in this case, recent exogenous change – and the notion that things are being transformed fast internationally owing to recent developments in technology, for which Europe must 'react' and 'keep up'. In this regard, global satellite navigation is framed as 'becoming central to all forms of transport and many other activities'. The second storyline pertains to uncertainty, with the Commission highlighting the need to 'ensure that European users are not at risk from changes in the service or excessive future charges or fees' as well as the 'threat to EU industry' because its own market for services could be 'undermined' (Commission 1998). The third storyline is about opportunity, with the idea that transport systems can be further integrated and new markets created domestically and in third countries, and that there can be human and economic benefits, including jobs, because '[global navigation satellite systems] can improve safety and efficiency in transport' and be cost-effective, with applications in a wide range of areas. The fourth storyline is about choice – do we [the EU] develop systems jointly or together? What system to choose? The fifth storyline is the solution (prognosis) or preferred course of action, as advocated by the Commission, to safeguard future social and economic welfare: 'Europe should develop a new satellite navigation constellation with appropriate terrestrial infrastructure (Galileo)'. The narrative is teleological, insisting that action must be taken now to protect the future.

Deep structure: ideologies

Looking beyond the specific discursive elements related to satellite navigation and Galileo, there appears to be four ideological traits underpinning the discussion and providing the political philosophical rational for the proposed course of action. First, the Member States have public obligations – that is, part of the state's role is to ensure the provision of transport infrastructure and services. Second, monopolies and dominant positions are bad and competition is good, which is to say that no global power should dominate in space policy and the provision of satellite services. Third, independence and control (over systems) are desirable, for political, economic and intellectual reasons, particularly when it comes to developing and imposing technological standards. Fourth, the EU should play a full role internationally. This may be an implicit assumption, but it is stated explicitly in the communication – that is, not only should the EU be seen to play a role on the global stage, but that the Member States should act together in this endeavour for the sake of 'credibility in negotiations', to ensure capacity and to 'have the confidence to invest'.

Frame dimensions over time

I have identified individual frames within Commission communications on three occasions at four-year intervals up until the end of the definition phase, as this corresponds largely with agenda-setting for Galileo, culminating in a political decision to back satellite navigation. In short, Galileo was 'defined' throughout the 1990s, leading up to the development and validation phase in the period 2001–2005. Although further analysis of the subsequent phases may identify further frames, it is assumed that they will be more static once a political decision has been taken. This explains the focus on issue definition and the frames at play in these formative policy stages.

Nascent frames

There were four Commissioners for Transport in this initial period in which Europe's space policy began to emerge: Karel van Miert (Belgium, 1989–1992), Abel Matutes (Spain, 1993–1994), Marcelino Oreja (Spain, 1994–1995) and Neil Kinnock (UK, 1995–1999). The Commission was operating in its 'golden period' under President Jacques Delors, in which it enjoyed considerable freedom on manoeuvre, first boosted legally by the Single European Act (1987), which sought to prepare for '1992' and the completion of the internal market. The Commission's scope for action was also increased by commitment at The Hague in 1987, but also by the ruling of the European Court of Justice in 1985, which found the Council guilty of failure to provide a Common Transport Policy, as set out in the Treaty of Rome (1958). The Commission's 1993 White Paper on *Growth, Jobs, Competitiveness: the Challenges and Ways Forward into the 21st Century* (Commission 1993) refers in its annex to the 'creation of infrastructures (cable and land or satellite-based radio communication), including integrated digital networks'. In its indicative list of trans-European networks, project 26 is stated as 'multimodal positioning system by satellites, for which studies are in progress by France, Germany and the ESA', while project 20 is 'air traffic management system for Europe', which 'includes also the satellite system Inmarsat-III (navigation payloads) and associated ground segment' (Commission 1993).

The period can be summed up as one of searching for policy coherence, identifying challenges, new approaches and potential applications, with an eye on the internal market, but also examining how European industry can expand externally. The very first communication from 1988 *The European Community and Space: A Coherent Approach* (Commission 1988) actually makes an explicit reference to 'framing policy'. It provided six action lines and reviewed all current national specialisations in the space domain. It sought to create some movement, implying that the Community had already reached a certain stage and was 'destined' for (fait accompli) further action. The Commission credited the expertise of the European Space Agency. The 1992 communication *The European Community and Space: Challenges, Opportunities and New Actions* recognised important achievements 'thanks to the ESA' and laid out a series of very broad Community objectives, advocating 'concentration and coordination' between Member States and relevant European organisations to identify 'common strategies, approaches and positions' (Commission 1992). The nascent frames operating in the early 1990s are summarised in Table 2.1.

Table 2.1 Nascent frames in Commission (1992) 360 final

Frames	F1: the state	F2: the market	F3: Europe and the world
Normative dimension (values; moral beliefs; why does the problem exist)	State intervention is desirable; space exploration is important for Europe	Competition, free market economy; space can be justly exploited for commercial gain	Europe is operating in an open, global, international system
Constitutive dimension (what is the problem; why intervene)	Governments under pressure to justify continued support for space programmes in terms of their potential economic importance	Increased competition is being felt in launch equipment, ground services and satellites	The multiplicity of space players and emerging new space-faring nations
Cognitive dimension (what has led to the problem; story about cause and effect relations)	Competitiveness of European space industry is fragile; Europe is small and fragmented; actions lie beyond the responsibility of space agencies	New demands are being made for space techniques, technologies and space-derived information; the Community must react	Vulnerability to increased global competition; absence of multilateral trade discipline; new relationships need to be forged
Policy dimension (what should be done; public action)	• Encourage and support development and exploitation of Earth observation applications • Ensure appropriate regulatory conditions to encourage the development of new markets in satellite communication services • Develop complementarity and synergy between Community R&TD programmes and space programmes of European Space Agency/Member States • Encourage the consolidation and growth of competitive space industry at international level, within framework of Community industrial and commercial policies • Encourage balanced international cooperation taking into account the former Soviet Republic and Central and Eastern Europe • Develop a trans-European satellite navigation system		

Evolving frames

Four years is a long time in the politics of agenda-setting and issue framing. In 1996, the Commission asserted that 'the overall international scenario has greatly evolved and new important events have occurred, requiring an update of the Commission's position in the space field' (Commission 1996). It stated, about itself: 'The Commission has already given impetus to this sector through a dedicated activity within the Fourth Framework Programme of RTD'. It framed its new communication in terms of space's role in society, economy and science over the last 30 years, documenting the creation of the European Space Agency alongside national agencies and institutions, which had 'allowed' the development of a wide range of technological capabilities and an important industrial infrastructure, in so doing 'putting Europe at the forefront of the space field'. The key messages here were that space should not be considered a narrow policy area, but one with commercial implications for several EU policy fields and, hence, for the Commission as a multi-organisation, of relevance to the remit of several Directorate-Generals, not just transport or industry. It is claimed there were 350 industrial companies in Europe with some degree of involvement in space activities, from the prime contractors to the first-, second- or third-level subcontractors and suppliers. Taking a wide definition, it was calculated that this sector generated up to €8 billion and represented over 30,000 industrial jobs, with a further 9000 in related institutions (such as space agencies, research centres, universities), compared with about 200,000 jobs in the space business in the USA.

The key impression of this follow-up communication from 1996 is that there has already been considerable progress and activity in the interim period, both in Europe and also elsewhere. The communication is imbued with a sense of things evolving quickly: 'space applications are developing fast'; 'applications-related industries and services which are, furthermore, growing at a faster rate than that of the space segment'; 'the markets for space applications are developing fast'; 'the overall international scenario affecting space is evolving at a great speed'; 'speedy approval of the action plan'; and 'to speed up the process'.

There are clues to future political tensions surrounding Galileo because the communication both places an emphasis on policy being 'at the service of the citizen' as well as the military. On the one hand it refers to the practical use of space technologies, such as monitoring land use and resources, coastal waters, oceans, the atmosphere, fisheries, major risks and hazards, but, on the other hand, it also contains explicit references to defence and dual-use space technologies. Although it is not within the Commission's remit to consider the military aspects of space technology applications, any European strategy should ensure the convergence of civil and military effort in order to avoid duplications and make the best use of the available public funding (Commission 1996). The report goes on to assert that:

> space technology is of central and growing importance for many types of military missions in the fields of telecommunications, navigation, intelligence,

early warning and meteorology' and that 'initially, military needs have driven space technology, which was channeled to civilian applications at an ulterior stage. More recently, the situation has changed...

(Commission 1996: 26).

The evolving frames at play in the mid-1990s are summarised in Table 2.2.

Mature frames

By the time the late Loyola de Palacio took over as Transport and Energy Commissioner in 1999, the issue frames with space policy discourse had already matured and specific courses of action and EU priorities had been hammered out. Following the 1998 communication *Towards a Trans-European Positioning and Navigation Network* and the 1999 communication *Galileo – Involving Europe in a New Generation of Satellite Navigation Services* (Commission 1998, 1999) came the 2000 communication *On Galileo* (Commission 2000). The Cologne European Council in 1999 and the Feira European Council in 2000 emphasised the strategic importance of Galileo and the need to decide whether to continue the programme.

The 2000 Commission Communication reminds Member States of the successive phases of the Galileo programme: the development and validation phase (2001–2005); the deployment phase (2006–2007); and the operating phase (2008 onwards), although, in hindsight, we know that there has been significant slippage in these timings. Unlike the previous documents, it pays much more attention to organisational features, such as the technical aspects of the systems architecture, managerial issues and financing. Ironically, it states 'cost/benefit studies show Galileo to be cost-effective and sufficiently attractive to obviate the need for any further public funding in the form of subsidies from 2007' – in 2007, three communications emerged in the light of the failed experiments with public–private partnerships, with the implementation of Galileo and the European Global Navigation Satellite System programmes 'at a cross-road' (Commission 2007a), the Commission advocating that Galileo should be re-profiled to ensure a positive outcome by turning its back on the notion of private funding sources (Commission 2007b) and trying to resurrect a politically lack-lustre European space policy in general (Commission 2007c).

It is important to note that, although satellite navigation was now high on the policy agenda – being discussed at European Council meetings – and with content largely defined, there was by no means political and financial commitment among the Member States to finance it. Hence, although the previously identified frames are found within the Commission (2000) communication, the emphasis is placed on change being upon the EU and for it being set to bring benefits – that is, change is: (1) favourable – a 'positive boost', 'added value', 'notching up success after success'; (2) considerable – 'satellite navigation revolution', 'leading-edge technology'; (3) rapid – 'from one moment to the next', 'coming up daily with new applications'; (4) essential – 'satellites are

Table 2.2 Evolving frames in Commission (1996) 617 final

Frames	F1: society/economy	F2: science/research	F3: trade/industry
Normative dimension (values, moral beliefs; why does the problem exist)	European society and the economy are interlinked; the EU works to improve quality of life for its citizens	European science and research/technology must be at the forefront internationally	European industry/trade are vital; the space industry is a key element in a 'much larger value added (service) chain'
Constitutive dimension (what is the problem; why intervene)	The emerging global information society brings rapid technological development; Europe risks falling behind; must keep up	Earth observation is emerging from the realm of science towards an increased role of the market; end of the geopolitical context that fuelled the growth of the space sector for decades	Markets for space applications are developing fast in telecommunications and navigation; competition for satellites and launch services is ever more intense/global despite economic slowdown
Cognitive dimension (what has led to the problem; story about cause and effect relations)	Inherent trans-border character of satellite services means opportunity for society and economy; impact on daily life	Space applications receiving increasing attention in the 4th Framework Programme for R&TD (1994–1998) and in EU policies, but insufficient	Some progress made restructuring European industry, but US competitors have concentrated efforts; Europe not spending as much as USA or Russia
Policy dimension (what should be done; public action)	• Avoid narrow concept of space industry • Space applications impact on many policy fields (such as information society, environment, agriculture, trans-European networks, regional policy, development aid) • EU should play role of 'pioneer customer' in space applications • High-level group of industrialists must define actions to promote dynamic and competitive space industry • Must develop a proper policy framework • Must improve conditions for industrial competitiveness		

indispensable for safeguarding independence', 'crucial link'; and (5) plural – 'a multitude of activities', 'many of them [possibilities] hitherto undreamed of'. Moreover, the communication features much 'stressing' and 'reaffirming', with things 'again should be borne in mind' (see Table 2.3).

What was clearly essential at this relatively advanced stage of agenda-setting was to secure maximum political support to proceed to the development and validation phase. The Commission 'reaffirms the strategic and economic importance of the project and endorses a proposal that it be continued beyond 2001'. The timing of the communication allowed the Commission to allude to 'a new Community approach adapted to a new century'; indeed, it was gearing up for its own White Paper *European Transport Policy for 2001: Time to Decide* (Commission 2001: 6–7).

To convince readers of Galileo's merits, it sought to convey the wide range of potential applications and to make clear the many policy areas that would benefit. It did so by offering simplified explanations and examples of how Galileo could be used – for example:

> in transport (positioning and measurement of the speed of moving bodies, insurance, etc.), to medicine (remote treatment of patients, etc.), law enforcement (surveillance of suspects, etc.), customs and excise operations (investigations on the ground, etc.), and agriculture (grain or pesticide dose adjustments depending on the terrain, etc.).
>
> (Commission 2001: 3)

Beyond the financial value of services that can be generated from satellite navigation applications, we can also sense that the Commission places a symbolic value on the EU possessing such technical mastery. The word 'accuracy' appears 13 times – for example, 'a high degree of accuracy', 'extreme accuracy', 'any degree of accuracy', 'degradation of signal accuracy', 'a greater degree of accuracy', 'positioning accuracy'. Other commonly used words include 'exact', 'precision' and 'pinpointing'; the word 'position' appears 38 times. We might consider that the Commission felt a promise of 'accuracy' in policy on the ground would have political salience for decision-makers.

Conclusion

This chapter has provided an analysis into how Galileo has been constructed discursively over time by the European Commission in its role as agenda-setter. It has shown that the way in which it has 'talked about' the need for satellite navigation has shifted over time. Thus to legitimise Galileo – in the face of competition from other policy areas for political and financial support – the Commission had to constantly adapt and refine its framing strategies. In terms of the European space programme, it shows how multiple and complementary frames were cultivated over time as part of a broader creative strategy to convince the various stakeholders in the EU Member States of the need for independent

Table 2.3 Mature frames in Commission (2000) 750 final

Frames	F1: environment	F2: independence	F3: interoperability
Normative dimension (values, moral beliefs; why does the problem exist)	Safety/security; management and monitoring	Political control; technological autonomy; ability to reorganise economy	Collaboration in European industry; additionality; cost-effectiveness; partnership
Constitutive dimension (what is the problem; why intervene)	Tighter controls are needed (transport); recent disasters (Erika); 'no infrastructure will be capable of managing its own growth unaided'; airports/airspace increasingly congested; geostationary satellites do not fully cover Europe's Nordic regions	'Economies more and more at the mercy of . . .'; 'can we really afford to be dependent on just one non-European signal in order to safeguard these turnovers and the jobs that will depend thereon'; 'the GPS constellation is completely outside Europe's control'	Lack of regulation/harmonisation; inter-relationships; need to synchronise signals and coordinate operating standards; potential advantages of EGNOS system 'hampered' by limitations
Cognitive dimension (what has led to the problem; story about cause and effect relations)	'Only Galileo will have the capabilities needed'; continuous coverage by a number of satellites of any point on the Earth's surface, including the polar regions	'A constellation of EU satellites is indispensable to safeguarding our independence'; Galileo will overcome limitations, provide new services	'Certain conditions are essential to the success of Galileo'; compatible technical standards; cost-effective solutions; system offers great added value to existing systems
Policy dimension (what should be done; public action)	Investors will participate under two conditions:		

- necessary political and financial decisions are taken to launch the next development and validation phases
- a single entity, operating within a precise institutional framework, is appointed to manage the programme

The EGNOS and Galileo infrastructures will have to be integrated and managed centrally competence in satellite navigation. Given the limited size of the EU budget (at just over 1 per cent of the total GNP of the Member States) and the problems of securing private finance, the difficulty in producing convincing frames to justify enormous expenditure on a European space programme should not be underestimated.

This example shows how 'sets of frames' operate over time and how institutions (authors) seek to persuade and convince by the way they present an issue to decision-makers in official documents. Frames are never static, but are dynamic and evolving. Issue frames relating to specific policies will depend on the larger frames used within the visionary (medium-term) documents of the EU, such as Treaties and White Papers. This analysis shows how frames evolve at the agenda-setting stage. Nascent frames create a sense of relevance and a rationale for initial political involvement, here drawing on broad ideologies about the role of the state in the market and the place for Europe vis-à-vis other policy actors. Having established a role for the Community, evolving frames create a sense of ongoing activity, evoking a journey that is underway and with reference to achievements so far, thus creating momentum and a forward trajectory. This is seen here by drawing attention to different sectors and policy actors in the economy, society, science and industry, who should all be active. With a reasonably high degree of issue definition – and perhaps conviction on the part of target groups – mature frames then present the issue in much more technical terms, making reference to practical uses, while the Commission keeps its eye on the persistent, most resonant issue frames of competitiveness/trade, safety/security and jobs/growth, and, in so doing, reaffirms the EU values and norms of cooperation, integration and cohesion. The Commission might purposefully seek to operate more than one frame – rather like running several horses in a race in the hope that one will be strong enough to reach the finish line – that is, secure sufficient political/financial support to result in a decision and, thereafter, implementation.

These frame sets are complementary over time rather than conflicting. Frame conflict may be more apparent in documents authored by more than one actor group, such as committees with competences in space policy, both across EU institutions and at the Member State level. To give another analogy, we might think of the multitude of frames that operate in one policy area by different institutional actors over time as being like a mirror ball. In this sense, each Commission Communication is like a patch of adjacent mirrors on that ball, alongside small mirrors (frames) perpetuated and projected by other voices (actors and institutions). Hence a multitude of frames spin on an issue axis, words going round and round, to reflect the issue in different ways, shining light on, sometimes even dazzling, policy-makers in the ballroom.

Acknowledgements

This chapter was first published as: Stephenson, P.J. (2012) 'Talking space: the European Commission's changing frames in defining Galileo', *Space Policy*, 28(2), 86–93.

The author is grateful to Elissaveta Radulova, whose approach to using frame analysis in her recent doctoral work, using tables for synthesis, inspired him in this analysis.

References

Note: While focusing in depth on three communications, the author identified and examined 29 relevant official documents published by the European Commission, which can easily be retrieved via internet searches and EUR-Lex, which provides free, direct access to EU legislation. These documents were: COM(93) 700, COM(88) 417 final, COM(92) 360 final, COM(94) 210, COM(96) 617 final, COM(97) 91 final, COM(98) 29 final, COM(99) 54 final, SEC(1999) 250, SEC(1999) 789, COM(2000) 750 final, SEC(2001) 1960, COM(2001) 370, COM(2002) 518 final, COM(2003) 123 final, COM(2003) 132 final, COM(2004) 112 final, COM(2005) 24, COM(2006) 272 final, COM(2007) 212, COM(2007) 261, COM(2007) 534 final, COM(2010) 308 final, COM(2010) 550, COM(2010) 700, COM(2011) 5/3, COM(2011) 398, COM(2011) 500, COM(2011) 676.

Amos, J. (2011) 'Europe's first Galileo satellites lift off' [online]. *BBC News*, 21 October 2011. Available from: www.bbc.co.uk/news/science-environment-15372540. Last accessed 30 April 2015.

Commission (1988) *The European Community and Space: A Coherent Approach. Communication from the Commission.* COM(88) 417 final, 26 July 1988.

Commission (1992) *The European Community and Space: Challenges, Opportunities and New Actions. Communication from the Commission to the Council and the European Parliament.* COM(92) 360 final, 23 September 1992.

Commission (1993) *Growth, Competitiveness, Employment: The Challenges and Ways Forward into the 21st Century – White Paper.* COM(93) 700 final/A and B, 5 December 1993.

Commission (1996) *The European Union and Space: Fostering Applications, Markets and Industrial Competitiveness. Communication from the Commission to the Council and the European Parliament.* COM(96) 617 final, 4 December 1996.

Commission (1998) *Towards a Trans-European Positioning and Navigation Network: a European Strategy for Global Navigation Satellite Systems (GNSS). Communication from the Commission to the Council and the European Parliament.* COM(98) 29 final, 29 January 1998.

Commission (1999) *Galileo. Involving Europe in a New Generation of Satellite Navigation Services. Communication from the Commission.* COM(99) 54 final, 10 February 1999.

Commission (2000) *On Galileo. Communication from the Commission to the Council and the European Parliament.* COM(2000) 750 final, 22 November 2000.

Commission (2001) *European Transport Policy for 2010: Time to Decide – White Paper.* COM(2001) 370 final, 12 September 2001.

Commission (2007a) *Galileo at a Cross-Road: the Implementation of the European GNSS Programmes. Communication from the Commission to the Council and the European Parliament.* COM(2007) 261 final, 16 May 2007.

Commission (2007b) *Progressing Galileo: Re-profiling the European GNSS Programmes. Communication from the Commission to the Council and the European Parliament.* COM(2007) 534 final, 19 September 2007.

Commission (2007c). *European Space Policy. Communication from the Commission to the Council and the European Parliament*. COM(2007) 212, 26 April 2007.

Commission (2011a) *A Budget for Europe 2020 – Part II. Communication from the Commission to the Council, the European Parliament, the Council, the European Economic and Social Committee and the Committee of the Regions*. COM(2011) 500 final, 29 June 2011.

Commission (2011b) *A Growth Package for Integrated European Infrastructures. Communication from the Commission to the European Parliament, the Council, the European Court of Justice, the Court of Auditors, the European Investment Bank, the European Economic and Social Committee and the Committee of the Regions*. COM(2011) 676, 19 October 2011.

Daviter, F. (2007) 'Policy framing in the European Union'. *Journal of European Public Policy*, 14(4): 654–666.

Dostal, J.M. (2004) 'Campaigning on expertise: how the OECD framed EU welfare and labour market policies – and why success could trigger failure'. *Journal of European Public Policy*, 11(3): 440–460.

Dudley, G. and Richardson, J. (1999) 'Competing advocacy coalitions and the process of "frame reflection": a longitudinal analysis of EU steel policy'. *Journal of European Public Policy*, 6(2): 225–248.

Dryzek, J. (1993) 'Policy analysis and planning: from science to argument'. In: F. Fischer and J. Forester (eds), *The Argumentative Turn in Policy Analysis and Planning*. London/Durham, NC: Duke University Press, 213–232.

Fischer, F. (1997) 'Locating frames in the discursive universe'. *Sociological Research Online*, 2(3). Available from: www.socresonline.org.uk/2/3/4.html. Last accessed 30 April 2015.

Fischer, F. (2003) *Reframing Public Policy: Discursive Politics and Deliberative Practices*. Oxford: Oxford University Press.

Goffman, E. (1986) *Frame Analysis. An Essay on the Organization of Experience*. Boston, MA: Northeastern University Press.

Kingdon, J.W. (2003) *Agendas, Alternatives and Public Policies*, 2nd edn. New York, NY: Longman.

Kohler-Koch, B. (2000) 'Framing: the bottleneck of constructing legitimate institutions'. *Journal of European Public Policy*, 7(4): 513–531.

Lynggaard, K. (2007) 'The institutional construction of a policy field: a discursive institutional perspective on change within the common agricultural policy'. *Journal of European Public Policy*, 14(2): 293–312.

Mörth, U. (2000) 'Competing frames in the European Commission – the case of the defence industry and equipment issue'. *Journal of European Public Policy*, 7(2): 173–189.

Mörth, U. (2007) 'Public and private partnerships as dilemmas between efficiency and democratic accountability: the case of Galileo'. *Journal of European Public Policy*, 29(5): 601–617.

Radulova, E. (2011) *Europeanization Through Framing? An Inquiry into the Influence of the Open Method of Coordination on Childcare Policy in the Netherlands*. Maastricht: Maastricht University.

Rein, M. and Schön, D.A. (1993) 'Reframing policy discourse'. In: F. Fischer and J. Forester (eds), *The Argumentative Turn in Policy Analysis and Planning*. London/Durham, NC: Duke University Press, 145–166.

Schön, D.A. and Rein, M. (1994) *Frame Reflection. Toward the Resolution of Intractable Policy Controversies*. New York: Basic Books/HarperCollins.

Smith, M. (2003) 'The framing of European foreign and security policy: towards a post-modern policy framework?' *Journal of European Public Policy*, 10(4): 556–575.

Snow, D.A. and Benford, R.D. (1988) 'Ideology, frame resonance, and participant mobilization'. In: B. Klandermans, H. Kriesi and D. Tarrow (eds), *International Social Movement Research: Volume 1*. London: JAI Press, 197–217.

Surel, Y. (2000) 'The role of cognitive and normative frames in policy-making'. *Journal of European Public Policy*, 7(4): 495–512.

Torfing, J. (2005) 'Discourse theory: achievements, arguments and challenges'. In: D. Howarth and J. Torfing (eds), *Discourse Theory in European Politics: Identity, Politics and Governance*. Basingstoke: Palgrave Macmillan, 1–32.

van Dijk, T.A. (1977) *Text and Context Explorations in the Semantics and Pragmatics of Discourse*. London: Longman.

Weaver, O. (2000) 'Discursive approaches'. In: A. Wiener and T. Diez (eds), *European Integration Theory*. Oxford: Oxford University Press, 197–215.

Part II
Polity
Institutional perspectives

3 The European Space Agency and the European Union[1]

Thomas C. Hörber

Introduction

The development of a sound relationship between the EU and the European Space Agency (ESA) has been the subject of much academic and political debate in recent years. This chapter reviews the debate and considers its origins and potential development trajectories. Although the process of European economic integration has enjoyed spectacular success since the 1950s (Camps 1964; Moravcsik 1998; Hörber 2006a), cooperation in the fields of space navigation, satellites, research and exploration has lagged well behind and was late to develop, although space fitted perfectly with Jean Monnet's definition of an ideal area for the advancement of European integration – that is, it was too big for individual nation states (Gaubert and Lebeau 2009: 41, 44), but was also a new field of politics, comparable with nuclear research under Euratom, which he strongly favoured (Hahn 1958: 1002; Gerbet 1999: 173).

After many false starts, ten European states[2] eventually agreed to the creation of the ESA in 1975, with the aim of cooperating in the field of space. The membership has since expanded to 19 European states, plus Canada (see Chapters 1 and 11).[3] There are, of course, significant differences between the EU and the ESA because the latter embraces non-EU states, such as Switzerland, and even extra-European states, such as Canada. There are also fundamental differences in the rules and procedures that make up the institutional soul of each organisation. The ESA is an archetypal old-style intergovernmental organisation of the conventional kind, with features such as a national veto in many areas, exclusively national funding and the controversial, but inevitable, concept of *juste retour*, whereby most national funding contributions effectively flow back to space companies in the same country (Gaubert and Lebeau 2009: 43). There is no element of supranational procedure, such as direct democratic legitimisation through an elected parliament, the EU's 'own resources' or qualified majority voting in the governing Council, although there is a simple or two-thirds majority voting in the ESA Council in some instances (see, for example, the ESA Convention, Article XI.5, available from the ESA website). The ESA was founded as an independent institution entirely separate from the European Communities, so the fact that there is now close cooperation with the EU is clearly significant and worthy of investigation. The German Minister for Research and Education, Edelgard Bulmahn, described developments in institutional collaboration and coordination (until 2005) as '… dovetailing the ESA into EU policy…' (FRG Ministry for Research and Education 2001) This seems to foreshadow, for the twenty-first century, the development of a common European space policy.

Research questions and methods

The main questions considered in this chapter are whether a common European position on space policy is developing and, if so, why is this happening now and what potential do these developments hold for the European integration process as a whole? This chapter approaches these questions through an analysis of past European collaboration in space affairs. It charts the recent process of closer involvement between the ESA and the EU and identifies the motivations underlying this process. It seeks to gauge the strategic potential for intensifying the coordination of national space efforts in the ESA and involving the EU. In the conclusion, the ever-closer relationship between the EU and the ESA is considered against the broader background of European politics and the continuing process of European integration.

Framing theory will be applied as the analytical structure to answer these questions. What is framing theory and how can it help to answer these research questions? Surel (2000) equates 'cognitive and normative frames' with what were once called 'paradigms' (Jenson 1989: 239, quoted in Surel 2000: 499). In some instances, frames also seem to border on ideologies, as when they are seen as endowed with the ability to make sense of the world, establish interaction in society and define the limits of social action (Surel 2000: 500). We could also call this 'identity creation', as occurs in the European integration process, although it is clear that the very definition of 'Europe' still varies greatly (Surel 2000: 507; Hörber 2006a). Space policy can be seen as an element in this identity creation process.

In a EU context, '… the move towards organized action is about an identity-seeking process' (Mörth 2000: 175). The connection to neoinstitutionalism, which seeks to explain the functioning of organisations, is made directly (March and Olsen 1996). The link to rational choice theories is also interesting in that framing theory assumes that individual actors make rational choices as to what they can actively influence – that is, concrete decisions in their lives – but that these decisions are based on pre-existing belief systems (Bachrach and Baratz 1962; Sabatier 1998: 109; Baumgartner 2001; Daviter 2012: 4–5). Daviter nuances it in this way:

> When the focus of attention shifts, some facets of a problem are emphasised or deemphasised, some aspects of a decision are revealed and others ignored. As the representation of the issue changes, so does the perception of what is at stake, and the preferred solutions vary in response
>
> (Daviter 2012: 6)

We might call this an underpinning ideology, a social paradigm or, in more recent literature, 'frames'. In the process of defining and revising frames, their ordering structures are reshaped – for example, the hierarchical rankings of values and norms, the resetting of interconnections between actors, institutions or policies (Surel 2000: 508; Smith 2003: 557, 571), or the process of redefining the relationship between the EU and the ESA. March (1997) called it 'sense-making', but we can also see ideas as a defining factor in this process, as put aptly by Goldstein and Keohane (1993):

> By ordering the world, ideas may shape agendas, which can profoundly shape outcomes. Insofar as ideas put blinders on people, reducing the number of conceivable options, they serve as invisible switchmen, not only by turning action onto certain tracks rather than others ... but also by obscuring the other tracks from the agent's view.
>
> (Goldstein and Keohane 1993: 12)

For a proper perception of the development of ideas, we need to consider their evolution over a long period of time (Jenkins-Smith and Sabatier 1993: 6; Dudley and Richardson 1999: 244; Sabatier and Jenkins-Smith 1999; Nilsson *et al.* 2009: 4456). For these researchers, ideas and advocacy coalitions are equated with policy 'frames' (Weiss 1989: 117; Dudley and Richardson 1999: 246; Daviter 2012: 2). The process of fitting the ESA into the European institutional framework is a very good example of the development of ideas and advocacy. In this process, supporters and opponents naturally take positions that are naturally different, and this leads to the formulation of policy (Weiss 1989: 656). The process is inclusive for all participants, which is where framing theory reveals its roots in discourse theory, which posits the hegemonic struggle over ideas/ policies as the key integrative concept in Western liberal democracies (Laclau and Mouffe 1985; Kohler-Koch 2000: 514, 527). New ideas can be positioned, policies can be reinforced or reformed, and institutions can be strengthened or questioned. 'Thus, frames concern power – the power to define and conceptualise' (Mörth 2000: 174). Importantly, however, this constructive power rather avoids conflict through displacement. Framing theory postulates that it is better to change the frame than to pursue the conflict within a previous frame. Taking a different perspective and adding new aspects to the discourse are seen as a constructive way forward in the political contest (Daviter 2012: 7). The creation of the European dimension in addition to national politics relating to Europe can be seen as an example to which future generations have been bound since its introduction. Parsons (2006) shows this aptly with the concept of path-dependency – that is, earlier frames delimit the choice available in later frames (Kohler-Koch 2000: 515). This is also the process of policy formulation in which frames compartmentalise politics so that this complex reality becomes manageable. Daviter (2012) calls this 'policy venues', but this is really the classical political science definition of 'policy' and thus 'frame' becomes synonymous with 'policy' in some literature references (Daviter 2012: 10). Evidently, the limits and contents of such frames change in the actualisation process of politics. The formulation and redefinition of a European policy for space is the example explored in this chapter.

Past European space projects: the ESA as a technical agency

Since its inception, the ESA has participated in numerous projects, missions and research programmes. It has cast itself as a technical agency, rich in expertise, but poor in political 'clout'. Access to space was one of the earliest examples of the development of a European space programme by the ESA (Gaubert and Lebeau 2009: 42; see also Chapter 1). The first major ESA project was the launch of a carrier rocket, Ariane, alongside other pioneering projects, notably Meteosat (a weather satellite), ECS (a telecommunication satellite) and MARECS (a maritime communication satellite). The first Ariane launch took place on 15 December 1979 from the launch pad at Kourou in French Guiana (Harvey 2003: 169). This venture established the European 'gateway' to space and broke the virtual monopoly enjoyed by the USA on commercial launches. In 1992, the launch pad was upgraded to accommodate the redesigned Ariane 5 (Harvey 2003: 190). The overall objective was to establish a commercial involvement in the Ariane programme and gradually hand over utility and control to commercial users, which has been successfully completed (FRG Ministry for Research and Education 2001). In recent years, the ESA has also developed the Vega launcher, which is a smaller and cheaper carrier, suitable for commercial applications in space. Vega – in collaboration with Soyuz launchers from the Russian space agency – is used to deliver Galileo satellites into orbit. There are also concepts for reusable launchers, which would greatly reduce the cost of launching satellites and spacecraft. Both developments are steps towards the commercialisation of space technology, which is seen by the European Commission as essential for the expansion of the European space sector (ESA 2004a: 1).

Another major ESA project has been ENVISAT 2002–2012 (Harvey 2003: 240), which is a good example of European engagement in earth observation – one of the most important fields of European space investment (FRG Ministry for Research and Education 2001: 3). ENVISAT has been equipped with a number of devices, including: (1) Medium Resolution Imaging Spectrometer (MERIS), which observes 'ocean colour' by means of the solar radiation reflected by the (open and coastal) ocean surface; (2) Michaelson Interferometer for Passive Atmospheric Sounding (MIPAS), which monitors chemical changes in the atmosphere; (3) Advanced Synthetic Aperture Radar (ASAR), designed to scan the oceans at night and through cloud; (4) Global Ozone Monitoring by Occultation of Stars (GOMOS), which monitors ozone depletion; (5) Radar Altimeter 2 (RA-2), which measures the ocean floor, waves, ice and the polar ice sheets; (6) Microwave Radiometer (MWR), which measures atmospheric humidity; (7) LRR, which gauges the distance between the earth and the satellite; (8) SCanning Imaging Absorption SpectroMeter for Atmospheric CHartographY

(SCIAMACHY), used to observe pollution in the atmosphere; (9) Advanced Along-Track Scanning Radiometer (AATSR), used to measure sea temperatures; and (10) Doppler Orbitography and Radiopositioning Integrated by Satellite (DORIS), which measures distances (Harvey, 2003: 242). The mass of data gathered by ENVISAT's suite of ten instruments now provides scientists with a global picture of our environment and is helping to meet the initial needs of the Global Monitoring for Environment and Security (GMES) initiative, pending the commissioning of the more sophisticated Sentinel satellites. Hence earth observation is one of the pillars of the overall strategy of establishing Europe as an information power with first-hand access to primary data about our planet.

In addition, a major project undertaken was Galileo, founded in July 2003. This has been hailed as marking the dawn of a new era in satellite navigation (Amos, 2011), which will provide a network for a precise timing and location service in competition with the USA's global positioning system. As far back as the early 1990s, the EU authorities agreed that Europe must have its own global navigation system. To implement the decision to increase European independence in space infrastructure, the EU and the ESA joined forces to develop Galileo as an independent programme under civilian control, guaranteed to operate at all times (see ESA website). Despite several serious setbacks (see Chapter 13), Galileo remains the most important joint EU–ESA venture. It has caught the imagination of the public and made international headlines, not just because it is considered to be the most important technology project in Europe, but also because of its political ramifications. Given the very close collaboration between the ESA and the EU, Galileo can truly be seen as a pan-European project with a common purpose. This is yet further evidence that a European space policy is now taking shape, in which Galileo and GMES[4] are major pillars (ESA 2005; Hörber 2006b: 19–28).

Finally, the International Space Station was built with substantial European participation from 1998. As with Ariane, this project is at a stage where the commercial use of facilities has become feasible and is, indeed, highly desirable (FRG Ministry for Research and Education, 2001: 3). The commercialisation of space technology is clear evidence that European space initiatives are reaching maturity. Increased private commercial activity will further increase the importance of space policies in European politics.

Well-publicised programmes such as the Mars Express and Huygens – an ESA probe sent to Titan and Saturn – have also increased public awareness of European space engagement. The political momentum for further cooperation and advance can clearly be felt. Similar impulses can be expected from Aurora, the ESA umbrella programme for space exploration (for more information, see the ESA website). The political repercussions are considerable, as currently reflected in the changes now mooted in ESA–EU institutional structures.

Different history, common future purpose? Framing a common destiny

The background of 'different history, common purpose' was described with much insight in the Council Resolution on a European Space Strategy in November 2000 (EC and ESA/C-M/CXL VIII/Res. I; see von der Dunk 2003: 83). Here, a frame of a common destiny for both the ESA and the EU was created. It recognised that the ESA was created in 1975 so that those European states with an interest in space could pool their resources to form a reputable space programme. In the ESA, with headquarters in Paris, an organisation was created that could provide more structure and a better focus than exclusively national projects (Crawford and Schulze 1990: 191). After a few bad experiences in relations with NASA before the foundation of the ESA, resulting in most instances from the frequent changes in US policies on technology sharing and trade, many European states found that they much preferred to work in a reliable European space organisation, rather than to remain dependent on assistance from the USA (Crawford and Schulze 1990: 191). This is one historical parallel between the ESA and the EU – the desire to reduce dependence on the USA, although cooperation between the EU–ESA and US institutions remains the closest by far on the international stage (for further information on ESA–USA cooperation, see ESA 2003a: 19).

Although the ESA is not part of the EU, the two organisations maintain close relations. The ESA and the EU seek to cooperate as much as possible to ensure Europe's access to space and cutting-edge research in the fields of satellites, communications, environmental monitoring and space technology. 2003 was a decisive year in the establishment of a common framework between the ESA and the EU. In January 2003, the European Research Commissioner, Philippe Busquin, introduced a Green Paper on a European space policy (Gaubert and Lebeau 2009: 42, footnote 4), aimed at launching the debate on Europe's space policy with all players – that is, national and international organisations, the European space industry, future users and the scientific community. For Europe's citizens in particular, the Green Paper was also designed to stimulate interest in European space engagement, which is an indication that European space affairs have achieved a prominence in government policies such that in the near future a more direct democratic mandate might be required than the ESA can provide. The involvement of the European Parliament seems to be the obvious choice – with direct democratic legitimacy as opposed to indirect legitimacy through ministerial representation in the ESA's Council of Ministers.

The Commission's Green Paper on space policy (Commission 2003a) was followed, in November, by a White Paper (Commission 2003b). In parallel, the EU and the ESA signed a Framework Agreement, which entered into force in May 2004 and proposed a structured framework for the relationship between the EU and the ESA (see ESA 2003a: 29–31, 2004a: 37). The Space Council was set up under this arrangement and met for the first time in November 2004. It consists of a joint meeting of the ESA Ministerial Council and the responsible

Council of the EU – that is, national ministers of research and development or economic affairs (see Chapter 5).

The agreement between the ESA and the EU recognised the specific complementary and mutually reinforcing strengths of the two bodies and committed them to working together, while avoiding unnecessary duplication of effort (ESA 2003b). Two main goals were identified. The first was to achieve progress towards a European space policy. This means that the EU will endeavour to meet the demands for services by using the ESA space programme and its infrastructure. In that respect, the ESA is acting in reality as an EU implementing agency. The second goal of the agreement was to make proper and suitable arrangements for cooperation between the two organisations, while recognising and respecting mutual independence. This is meant to facilitate joint space activities and provide a stable framework for EU–ESA cooperation. The objectives are ambitious and could open the door to new ways of cooperation, such as ESA management of EU space activities and EU participation in ESA projects (Creola 2001: 87).

On 7 June 2005, the Space Council decided on the sharing of roles and responsibilities at the highest level, and established priorities and guidelines. Accordingly, the EU is in charge of ensuring the exploitation of space for the benefit of its citizens, coordinating requirements and securing the coordination and promotion of a single European position on the international stage. This means that the EU has a framework-setting function and ensures the representation of European space interests abroad. The ESA and its Member States are in charge of space exploration and space science, and provide the tools needed for space activities, in particular actual access to space and the necessary technology. In Galileo and GMES, for example, the priority for space applications to benefit Europe's citizens has been spelled out. The ESA will continue to manage such programmes and cater more for the practical side of space technology, although the dividing line with the political side, which may be seen as an EU responsibility, is by no means as clear-cut (Hobe 2004: 27).

The possibility of an EU space programme that would absorb the ESA is also under discussion. There are pros and cons for the incorporation of the ESA into the EU – for the opposite proposals of EU membership in the ESA, see Gaubert and Lebeau (2009: 43) and von der Dunk (2003: 85). The main considerations are that the EU has, as its vocation, the representation of the best interests of the European people and it could therefore reasonably claim that the eminently important area of space activities should come under EU auspices for this very reason. In that way the EU could provide its citizens with additional benefits. The EU and its Member States have frequently been concerned that concrete and immediate benefits from space investment should be delivered to European citizens (FRG Ministry for Research and Education 2001: 1). On the other hand, Euro-sceptics argue that the reason why the ESA has been moderately successful is precisely because it is *not* under EU management and that the EU administration is already bloated and would therefore be unable to manage a space programme properly (Crawford and Schulze 1990: 144; von der Dunk 2003: 85;

Gaubert and Lebeau 2009: 37); for a discussion of successful (intergovernmental) projects such as CERN, ECMWF, ESO and EMBO, see Gaubert and Lebeau (2009: 38). The establishment of the Space Council with the ESA–EC Framework Agreement in 2003, and the more recent creation of a Directorate-General for Transport and Space at the Commission, highlight the growing importance of space policies. However, the power play between the Commission and ESA Member States also clearly reflects the declining importance of the Space Council (see Chapter 5) and the yet unresolved role of the Commission in space policies.

In summary, the debate about changing the institutional framework of the ESA and the EU to arrive at a common or, at least, a more internally consistent space policy, is a very strong indication that the superlatives in government statements – see, for example, the German Research and Education Minister's statement at the ESA Ministerial Council, Edinburgh, 14 November, 2001 (FRG Ministry for Research and Education 2001) – and press releases are no mere exaggeration. Space policies have acquired greater prominence in Europe over the last decade and this is the background to the intensifying debate about the right relationship between the ESA and the EU.

It almost goes without saying that space sciences are seen as a crucial aspect of the efforts to make Europe fit for the twenty-first century. The industrial application of scientific results from pure research – for example, on the International Space Station – is one major objective. Information networks in communication satellites and information gathering in earth observation, such as GMES, are just as important for a successful European future. In addition, an EU space programme would be a vital component in a common European defence strategy (Bildt and Peyrelevade 2000: 6). Sensitive information currently used under the auspices of the European Security and Defence Policy depends to a considerable extent on a sophisticated space infrastructure (von der Dunk 2003: 84). If the EU is serious about establishing satellite military intelligence-gathering capabilities – for example, under the Western EU (WEU Council of Ministers 1998), it will need a sophisticated space programme to support them. The ESA seems to be perfectly suited for the operative management of such space infrastructure (see Chapter 7). The connection between the European Security and Defence Policy and space can be seen in the inclusion of extra-European partners, such as Russia, in the European space programme where neighbourhood policies have been established (Smith 2003: 563). European frames have been exported and have been accepted by external partners in the space sector to give sense to concrete activities. Smith (2003) calls this a postmodern and post-sovereign European foreign policy, thereby adding another dimension to still existing national foreign policies (Smith 2003: 569, 570).

The White Paper also made recommendations for the future relationship between the ESA and the EU: '...it makes sense to aim for a closer institutional integration, thus ensuring the place of space issues in the overall evolution of European policies' (Bildt and Peyrelevade 2000: 7). The ministers responsible for space affairs agreed on the need for '...a process of institutional convergence that does not exclude bringing the present ESA within the Treaty framework of the European Union' (Bildt and Peyrelevade 2000: 7). The ministers proposed that the European Council should define a policy for space every five years. The ESA should include defence strategies (Slijper 2008) and there should be opportunities for debate in the European Parliament on such space matters (Bildt and Peyrelevade 2000: 7; for reference to the European Parliament, see Chapter 4). Again, as space affairs have steadily become more important in European political considerations, the issue of democratic legitimacy has become increasingly important. Their increasing centrality should be reflected in political legitimacy and eventually financial sanction from the European Parliament, which again is a strong argument for bringing the ESA into the political framework of the EU. However, this is not as simple as it seems, as Gaubert (a former Secretary General of Eurospace) and Lebeau (2009) point out:

> Dealing with it [harmonisation of the roles of the ESA and the EU] by saying that the ESA must become an agency of the Community would be like imagining the problem can be 'magicked' away. The crux of the problem consists of establishing relations between two entities of profoundly different character so that the ESA can become the executive arm of Brussels in space matters without losing its own dynamism.
>
> (Gaubert and Lebeau 2009: 43)

Strategic potential: the foundations for a stronger frame

For the EU, the strategic potential of space is clearly very considerable. There are many possible future fields of engagement, such as commercial launches, missions to Mars, Moon research or a Moon base, the observation of Venus and the completion of the Galileo satellite system – a summary of objectives was outlined by the second Space Council on 7 June 2005 (ESA 2005: 1). As outlined by the then European Commissioner Günther Verheugen, 'Space is an area where the added value of a joint and coherent policy on the European level is very clear. The industrial dimension of space is key to increasing the competitiveness of European industry' (ESA 2004b: 1; see also Verheugen 2005). Thus it is clear that the industrial and economic potential of this area of activity is now fully appreciated by the European political authorities. Consequently, there is further motivation to develop a European space policy politically through strengthening the common frame between the ESA and the EU.

Pressure for the militarisation of the EU has found opposition in a reassertion of the EU as a 'civilian power' (Yakemtchouk 2005; Telò 2006: 51). The case for larger military budgets for European countries has been made (Salmon and Shepard 2003: 206). Others argue that Europe is not seriously considering becoming a military power and ought not to do so (Telò 2006: 54, 145). Europe woefully lacks military capabilities, but must be seen as a very effective and indeed powerful international actor because it has shown that it is perfectly able to turn to good use its unquestioned and fully demonstrated abilities as a civilian power (Telò 2006: 57). Again, in contrast with the inclination of the USA to deploy the military as a direct power tool, the EU has agreed (under the Petersberg tasks) to increased military capabilities only as a means of making its civilian engagement more credible – that is, military intervention as a means of last resort – the existence of which might make opponents more susceptible to the preceding peaceful exercise of influence (Telò 2006: 75). Telò also stresses alternative and more innovative avenues of future development and cooperation than the military – for example, space endeavours such as the Ariane and the Galileo projects.

In its very European way, Galileo, unlike the US global positioning system, is not intended primarily for military use, but for civilian use only (Telò 2006: 54, 176). Thus a European space policy could become another component of European 'soft power' expertise and be deployed, alongside other elements, as an effective tool in the field of foreign policy. An active European space policy could therefore provide a reasonable alternative to the (military) power politics of other major powers in the world, an approach that would be more in line with the EU's history of anti-war development and, indeed, the foundation of ESA as an exclusively civilian institution. If it is right to believe that sophisticated information technologies will be central to political authority in the future, then space technology could help to provide the European authorities with this currency. This specifically European path need not lack power or influence.

Conclusion: competition or common purpose?

From a political perspective, it is the narrowing gap between aspiration and what is feasible in space that makes space policies so intriguing. It is the combination of potential for profit, with all the positive repercussions for a competition-driven economy, and the still largely untapped potential of increasing popular interest in space, which could eventually generate political support for grander space projects in the future. This is the strategic and political background to the debate concerning the relationship between the ESA and the EU. On the one hand, the EU has always been an eminently political organisation sensitive to the political potential any innovation may offer. A European space policy is only one of the most recent examples. On the other hand, the ESA has outgrown its technical roots and has acquired a political relevance that goes beyond merely administering limited national investments in space (Hobe 2004: 27).

Hence the idea of bringing the ESA under the EU roof would be another political step that could well serve to further enhance the political influence of both institutions in the future. It would also confer on space policies a political status that could well be an indication of the importance this field will have in the future. An indication of the increasing centrality of space policy was given as early as 2001 (FRG Ministry for Research and Education 2001: 2). Further appreciation of the future importance of a European space policy was expressed at the second Space Council meeting held in Luxembourg (ESA 2005) and the debate around Galileo (see Chapter 13) is clear evidence of the growing importance of space in

European politics. A concerted and common European space policy seems to be the next logical step, taking Europe forward to a position long enjoyed by the USA (Hobe 2004: 27). This would implement one element of further integration of all of Europe's public organisations (Gaubert and Lebeau, 2009: 44). It has been argued in the introduction to this book that space policies may serve as a spur to European integration and as an inspiration to its citizens. Such a European space policy may even have a considerable impact on strengthening European identity (Peter and Stoffl 2009: 36; Venet and Baranas 2012). Finding a stable equilibrium between the ESA and the EU will clearly be decisive if European space polices are to develop their full potential for European integration.

Notes

1 This chapter is based on a paper originally published in 2009: Hoerber, T.C. (2009) 'The European Space Agency (ESA) and the European Union (EU) – the next step on the road to the stars'. *Journal of Contemporary European Research*, 5(3): 405–414.
2 The founding members of the ESA were: Belgium, Denmark, France, Germany, Italy, Netherlands, Spain, Sweden, Switzerland and the UK.
3 Austria, Finland, Greece, Ireland, Luxembourg, Norway, Portugal and the Czech Republic joined later. Since 1 January 1979, Canada has had the special status of ESA 'cooperating state'.
4 For GMES, the Frascati Agreement, signed on 26 October 2005, provides for ESA–EU cooperation in earth observation.

References

Amos, J. (2011) 'Europe's first Galileo satellites lift off' [online]. *BBC News*, 21 October 2011. Available from: www.bbc.co.uk/news/science-environment-15372540. Last accessed 30 April 2015.

Bachrach, P. and Baratz, M. (1962) 'The two faces of power'. *American Political Science Review*, 56(4): 947–952.

Baumgartner, F.R. (2001) 'Political agendas'. In: N. Polsby (ed.), *International Encyclopaedia of Social and Behavioural Sciences: Political Science*. New York, NY: Elsevier.

Bildt, C. and Peyrelevade, J. (2000) *Towards a Space Agency for the European Union*. Paris: ESA Publications Division.

Camps, M. (1964) *Britain and the European Community, 1955–63*. London/Princeton, NJ: Princeton University Press.

Commission (2003a) *Green Paper. European Space Policy*. COM(2003) 17 final, 21 January 2003.

Commission (2003b) *White Paper. Space: a New European Frontier for an Expanding Union. An Action Plan for Implementing the European Space Policy*. COM(2003) 673, 11 November 2003.

Crawford, B. and Schulze, P.W. (1990) *The New Europe Asserts Itself: a Changing Role in International Relations*. Berkeley, CA: University of California Press.

Creola, P. (2001) 'Some comments on the ESA/EU space strategy'. *Space Policy*, 17(2): 87–90.

Daviter, F. (2012) *Framing Biotechnology Policy in the European Union*. ARENA Working Paper 05/2012. Oslo: ARENA Centre for European Studies.

Dudley, G. and Richardson, J. (1999) 'Competing advocacy coalitions and the process of frame reflection: a longitudinal analysis of EU steel policy'. *Journal of European Public Policy*, 6(2): 225–248.

Dunk, F. von der (2003) 'Towards one captain on the European spaceship – why the EU should join ESA'. *Space Policy*, 19(2), 83–86.

European Space Agency (2003a) *Annual Report 2003*. Paris: ESA Publications Division.

European Space Agency (2003b) *Press Release 76–2003, 12 November 2003*. Paris: ESA Publications Division.

European Space Agency (2004a) *Annual Report 2004*. Paris: ESA Publications Division.

European Space Agency (2004b) *Press Release 62–2004, 29 November 2004*. Paris: ESA Publications Division.

European Space Agency (2005) *Press Release 29–2005, 7 June 2005*. Paris: ESA Publications Division.

FRG Ministry for Research and Education (2001) *Press Release 180/2001, 14 November 2001*.

Gaubert, A. and Lebeau, A. (2009) 'Reforming European space governance'. *Space Policy*, 25(1): 37–44.

Gerbet, P. (1999) *La Construction de l'Europe*, 3rd edn. Paris: Imprimerie Nationale Éditions.

Goldstein, J. and Keohane, R. (eds) (1993) *Ideas and Foreign Policy: Beliefs, Institutions and Political Change*. Ithaca, NY: Cornell University Press, 3–30.

Hahn, H.J. (1958) 'Euratom: the conception of an international personality'. *Harvard Law Review*, 71(6): 1001–1056.

Harvey, B. (2003) *Europe's Space Programme: To Ariane and Beyond*. Chichester: Praxis.

Hobe, S. (2004) 'Prospects for a European space administration'. *Space Policy*, 20(1): 25–29.

Hörber, T. (2006a) *The Foundations of Europe – European Integration Ideas in France, Germany and Britain in the 1950s*. Wiesbaden: Springer Science and Business Media.

Hörber, T. (2006b) 'To boldly go where no one has gone before: the development of a European Space Strategy (ESS)'. *L'Europe en Formation*, 2/2006, 19–28.

Jenkins-Smith, H.C. and Sabatier, P.A. (1993) 'The study of public policy processes'. In: P. Sabatier and H.C. Jenkins-Smith (eds), *Policy Change and Learning*. Boulder, CO: Westview Press, 1–9.

Jenson, J. (1989) 'Paradigms and political discourse: protective legislation in France and the United States before 1914'. *Canadian Journal of Political Science*, 22(2): 235–258.

Kohler-Koch, B. (2000) 'Framing: the bottleneck of constructing legitimate institutions'. *Journal of European Public Policy*, 7(4): 513–531.

Laclau, E. and Mouffe, C. (1985) *Hegemony and Socialist Strategy*. London: Verso.

March, J. (1997) 'Understanding how decisions happen in organizations'. In: Z. Shapira (ed.), *Organizational Decision Making*. Cambridge: Cambridge University Press, 9–34.

March, J. and Olsen, J. (1996) 'Institutional perspectives on political institutions'. *Governance*, 9(3): 247–265.

Mörth, U. (2000) 'Competing frames in the European Commission – the case of the defence industry and equipment issue'. *Journal of European Public Policy*, 7(2): 173–189.

Moravcsik, A. (1998) *Choice for Europe: Social Purpose and State Power from Messina to Maastricht*. London: UCL Press.

Nilsson, M., Nilsson, L. and Ericsson, K. (2009) 'The rise and fall of GO trading in European renewable energy policy: the role of advocacy and policy framing'. *Energy Policy*, 37(4): 4454–4462.

Parsons, C. (2006) *A Certain Idea of Europe*. Ithaca, NY: Cornell University Press.

Peter, N. and Stoffl, K. (2009) 'Global space exploration 2025: Europe's perspectives for partnerships'. *Space Policy*, 25(1): 29–36.

Sabatier, P. (1998) 'The advocacy coalition framework: revisions and relevance for Europe'. *Journal of European Public Policy*, 5(1): 98–130.

Sabatier, P. and Jenkins-Smith, H.C. (1999) 'The advocacy coalition framework: an assessment'. In: P. Sabatier (ed.), *Theories of the Policy Process*. Boulder, CO: Westview Press, 117–166.

Salmon, T.C. and Shepard, A.J.K. (2003) *Towards a European Army – A Military Power in the Making?* Boulder: Lynne Rienner.

Slijper, T. (2008) *From Venus to Mars*. Amsterdam: Transnational Institute.

Smith, M. (2003) 'The framing of European foreign and security policy: towards a postmodern policy framework?' *Journal of European Public Policy*, 10(4): 556–575.

Surel, Y. (2000) 'The role of cognitive and normative frames in policy-making'. *Journal of European Public Policy*, 7(4): 495–512.

Telò, M. (2006) *Europe: A Civilian Power? European Union, Global Governance, World Order*. Basingstoke: Palgrave Macmillan.

Venet C. and Baranes, B. (eds) (2012) *European Identity through Space – Space Activities and Programmes as a Tool to Reinvigorate the European Identity*. Vienna: Springer.

Verheugen, G. (2005) 'Europe's space plans and opportunities for cooperation'. *Space Policy*, 21(2): 93–95.

Weiss, J.A (1989) 'The powers of problem definition: the case of government paperwork'. *Policy Sciences*, 22(2): 97–121.

Western European Union Council of Ministers (1998) *Rome Declaration, 16–17 November 1998* [online]. Available from: www.weu.int/documents/981116en.pdf. Last accessed 2 May 2015.

Yakemtchouk, R. (2005) *La politique étrangère de l'Union Européenne*. Paris: Harmattan.

4 Europe in space

The European Parliament's justification arsenal

Emmanuel Sigalas

Introduction[1]

The modern EU, featuring exclusive competences in five policy areas, shared competences in 11 areas and supporting or coordinating competences in more than ten policy areas, is a far cry from the European Communities (EC) of the 1950s with their comparatively modest portfolio (Nugent 2010). Few people back then would have believed that one day the Member States would pool part of their national sovereignty to create a common European policy on space. Yet this is what happened when the Member States ratified the Lisbon Treaty in 2010 and the EU was formally granted rights to 'define and implement programmes' related to space (European Union 2010: TFEU Art 4: 3).

This chapter aims to shed some light on how a European space policy (ESP) became a legitimate EU policy area. To do this, I looked at European Parliament (EP) Resolutions on space and analysed their content while bearing in mind the following question: why should action be taken in the field of space policy? Exploring the formal reasoning retrieved from the EP Resolutions revealed how the incremental expansion of the EU's competences in space affairs came to be portrayed as an acceptable or even inescapable choice.

The working assumption behind this study was that it matters how a particular topic is 'framed'. As Chong and Druckman (2007a: 104) explained, 'an issue can be viewed from a variety of perspectives and be construed as having implications for multiple values or consideration'. Similarly, Entman (1993: 55) argued that '[f]rames call attention to some aspects of reality while obscuring other elements, which might lead audiences to have different reactions'.

A number of studies have confirmed the importance of framing in influencing public opinion (e.g. Druckmann 2001; Chong and Druckman 2007b; Valkenburg *et al.* 1999) or EU policy choices (e.g. Littoz-Monnet 2014; Nilsson *et al.* 2009). Therefore I did not examine whether the EP's proposals were eventually adopted or not, which means that I did not test whether the EP's framing choices made any difference in policy-making. Instead, I identified the key frames or themes that the EP has deployed over the years to justify and legitimise the steps that eventually led to the ESP as we know it today.

I hypothesised that the EP chose to adopt and highlight a particular set of values, priorities and arguments in its Resolutions because it thought that this particular set would resonate best among its different audiences. Even though the research design does not allow for formal hypothesis testing, the data pattern leaves little doubt that the EP did not choose its justifying argumentation randomly, but did so strategically.

In the following section I explain why the EP Resolutions are a valuable source of information for studying the development of the ESP. Following that, I present the methodology I relied on to condense and analyse the content of the 22 EP Resolutions on space to date. The empirical findings sections present the EP's justification arsenal and how the EP used the latter strategically over the years to convince its audiences that Europe should have a place in space.

Public justification and the EP Resolutions on space

Space-related endeavours are politically sensitive. They are expensive, high risk and potentially useful to the military. Hence the gradual involvement of the EU in space affairs was and remains controversial. In addition, the pursuit of an ESP is by no means self-explanatory. The resources of the EU and the Member States are limited and it is not immediately obvious why public money should be directed to space, given the cross-policy competition for attention and resources. Furthermore, cross-national interests in space are unequally divided between the EU Member States. Although some European states may have a lot to benefit from supra-nationalising space efforts (and expenses), others have few incentives. Consequently, the EU institutions need to be in a position to justify, first, the necessity of an ESP and, second, the evolving ESP objectives, resources and outcomes. They need to justify the ESP not only to each other, as they are all involved in the EU decision-making process, but also to external actors whose support is necessary (e.g. the ESA) and, last but certainly not least, to the broader public, who will foot the bill and cast their vote in national and European elections. Convincing the latter audience, the European voters, is particularly important. Politicians are more likely to pursue a policy if they anticipate that their voters will reward them and their party electorally. Thus the EP and the Council are more likely to agree on a policy that will resonate with their voters back home, whereas the Commission is more likely to initiate legislation if it believes that the two co-legislating bodies will agree to it.

The public justification of the ESP is therefore not only politically useful, but indispensable. It legitimates the ESP by openly providing reasons and explanations as to why it is needed, or why it ought to develop in a particular way. But what are these legitimating arguments that help to present the ESP in a positive light, attracting both public and inter-institutional support? Previous studies offer little guidance because, first, the ESP is a relatively new policy area and there is hardly any literature and, second, because research on the legitimation of other EU policies has not systematically studied the public argumentation of the EU institutions. This chapter aims to cover this gap.

Of the three main EU institutions, it is the two supranational ones, the Commission and the EP, that have the greatest incentive to promote their policy remit and consolidate an ESP. It is certainly worth studying the justification arguments of both, but at the same time there are compelling reasons to prioritise the examination of the EP's public discourse. The EP is the only directly elected EU institution representing European citizens. Thus it is its duty to justify its choices and position publicly, which means that it is the natural venue to explain why an ESP with a given set of objectives is needed.

The best place to look for the EP's justification for a common space policy is its Resolutions. The EP Resolutions are formal documents in which the official position of the EP is reflected. They are legally non-binding texts; this means that although they do not directly impact on EU legislation, they highlight what the EP believes ought to be done or avoided on a given matter. As a result, the policy positions found in them may deviate from the formal explanations presented in the EU legislation documents, but they are neither irrelevant nor inconsequential. Since the Lisbon Treaty came into effect, space has fallen under the ordinary legislative procedure that gives the EP co-decision rights. Consequently, space-related legislation cannot be passed without the consent of the EP and this automatically makes it a policy actor whose views count. Furthermore, the EP exercised some influence on the development of the ESP, even before the latter was recognised as a distinct EU policy area in the Lisbon Treaty (Sigalas 2012). Thus it is worth studying not only the post-Lisbon EP Resolutions, but also the earlier Resolutions.

Unlike the EP legislative reports, where the EP simply responds to the proposal tabled by the Commission, the Resolutions leave room for the EP to express its opinions without worrying whether other institutions will agree to them. Although sometimes the EP Resolutions are triggered by a Communication published by the Commission, they are often own-initiative reports. In any case, the EP is not obliged to adopt the argumentation or the views of other institutions when drafting its Resolutions. As a result, these Resolutions are ideal sources with which to study the genuine views and intentions of the EP in relation to the ESP.

Data and methodology

The EP started issuing Resolutions on space-related topics well before the signing of the Lisbon Treaty. The earliest Resolution I found dates back to 1979. The latest Resolution I included in my analysis was adopted in 2013. In between these dates, the EP issued 25 Resolutions, which brings the total number to 27 documents (Table 4.1). I analysed the content of 22 of these, omitting from the analysis only the highly technical texts. The first thing to be noticed when studying the EP Resolutions on space is that their length, complexity and sophistication grew over time. As Figure 4.1 shows, the 1979 Resolution was barely one page long, whereas the 2012 Resolution, for example, is ten pages long with more than 65 paragraphs.

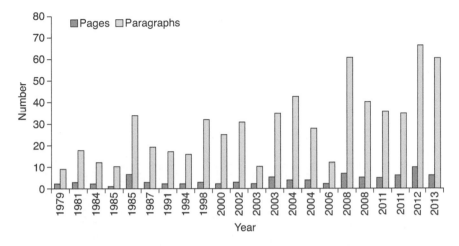

Figure 4.1 Length of the European Parliament's Resolutions on space policy (source: author's own calculations based on data about the Resolutions (see Table 4.2)).

Such a growth over time is to be expected and reflects two things. First, the growing interest and engagement of the EU in space affairs and, second, the growing number and complexity of the space-related topics that the EP feels need to be addressed. In short, even without reading the content of the Resolutions, it becomes obvious that, over the years, the EP has been attaching increasingly more importance to space as a European level policy area.

The total word length of the 22 Resolutions on space over a time span of nearly 35 years poses a formidable challenge in identifying the ESP legitimation frames. Nevertheless, using the 'why should action be taken' question as a selection criterion, I managed to detect and hand-code 224 justification arguments. The arguments may be a whole sentence or a quasi-sentence – that is, a fraction of a sentence that contains a unique justification argument.

Through an inductive process of looking for conceptual similarities between arguments, I identified 19 justification frames and two super-frames, which are presented in some detail in the following sections. In some cases it was possible to ascribe more than one frame to a given argument, but for reasons of simplicity and visual clarity I am presenting here only the most important frames.

Hand-coding inevitably raises inter-coder reliability issues (Krippendorff 2004). In the present context it means that a different coder or a text-reading machine might have identified different frames, more or fewer in number. Although it has not been possible to control for inter-coder reliability, any errors are likely to be too small to alter the interpretation of the findings. Automated coding, on the other hand, cannot substitute for human judgement, which is necessary in exploratory research when case-specific classification categories are determined inductively (Grimmer and Stewart 2013). Consequently, the legitimation frames presented

Table 4.1 List of European Parliament Resolutions on space policy

Date	Reference no.	Title
21 May 1979	OJ C 127/42 PE 56.322/final	Resolution on Community participation in space research
12 October 1981	OJ C 260/102	Resolution on European space policy
15 October 1984*	PE 95.639/final	Motion for a Resolution on the preparation of a draft Treaty on the ownership of space and peaceful exploitation of the resources of space
18 October 1984	PE 95.639/final	Motion for a Resolution on European space policy
5 November 1984*	PE 95.639/final	Motion for a Resolution on satellite remote sensing and world development
9 November 1984*	PE 95.639/final	Motion for a Resolution on a European space laboratory
1 February 1985	PE 95.639/final	Motion for a Resolution on establishing a European space policy including a manned European space shuttle and space resolution
30 September 1985	PE 95.639/final	Report drawn on behalf of the Committee on Energy and Research on European space policy
20 July 1987	OJ C 190/78	Resolution on European space policy
22 October 1991	OJ C 305/26	Resolution on European space policy
19 January 1993*	A3–0344/92	Resolution on a common approach in the field of satellite communications in the European Community
25 July 1994	OJ C 205/467	Resolution on Community and space
6 March 1995*	OJ C 056/180	Resolution on the communication from the Commission to the Council and the European Parliament on satellite communications: the provision of – and access to – space segment capacity
13 January 1998	A4–0384/1997	Resolution on the Commission communication to the Council and the European Parliament 'The European Union and space: fostering applications, markets and industrial competitiveness' (COM (96)0617-C4–0042/97)
18 May 2000	A5–0119/2000	Resolution on the communication of the Commission on the Commission working document 'Towards a coherent European approach for space' (SEC (1999) 789-C5–0336/1000–1999/2213(COS))
17 January 2002	P5_TA(2002)0015 A5–04541/2001	Resolution on the Commission communication to the Council and the European Parliament on Europe and space: turning to a new chapter (COM (2000) 597-C5–0146/2001–2001/2072(COS))
17 March 2004	OJ C 67 E/279 P5_TA(2003)0217	Resolution on the European Space Agency

Date	OJ reference	Resolution
31 March 2004	OJ C 81 E/90 P5_TA(2003)0427	Resolution on European Space Policy – Green Paper (2003/2092(INI))
21 April 2004	OJ C 96 E/128 P5_TA(2004)0051	Resolution on the communication from the Commission to the European Parliament and the Council on the state of progress of the Galileo programme (COM(2002)518–2003/2041(INI))
24 April 2004	OJ C 96 E/136 P5_TA(2004)0054	Resolution on the action plan for implementing the European space policy
15 December 2006	OJ C 306 E/393 P5_TA(2006)0385	Resolution on taking stock of the Galileo programme
3 December 2009	OJ C 294 E/69 P6_TA(2008)0365	Resolution of 10 July 2008 on space and security (2008/2030(INI))
22 January 2010	OJ C 16 E/57 P6_TA(2008)0564	Resolution of 20 November 2008 on the European space policy: how to bring space down to earth
11 December 2012	OJ C 380 E/84 P7_TA(2011)0265	Resolution of 8 June 2011 on the mid-term review of the European satellite navigation programmes: implementation assessment, future challenges, and financing perspectives (2009/2226(INI))
11 December 2012	OJ C 380 E/1 P7_TA(2011)0250	Resolution of 7 June 2011 on transport applications of Global Navigation Satellite Systems – short- and medium-term EU policy (2010/2208(INI))
6 August 2013	OJ C 227 E/16 P7_TA(2012)0013	Resolution of 19 January 2012 on a space strategy for the European Union that benefits its citizens (2011/2148(INI))
10 December 2013	P7_TA-PROV(2013)0534	Resolution of 10 December 2013 on EU space industrial policy, releasing the potential for growth in the space sector (2013/2092(INI))

Note
* Not included in the analysis. The year of the publication in the *Official Journal* (OJ) in some cases differs from the year the Resolution was first published as a European Parliament document.

here may not be the only possible constellation, but are the obvious constellation when viewing the Resolutions through the prism of why the EP believes space is a legitimate area of European involvement.

The utilitarian justification strategy of the EP

Having read all the EP Resolutions on space, from 1979 to 2013, the first thing one notices is how favourable the EP has always been to the idea of Europe as a whole playing a greater role in space. What is striking is not so much that the EP has been supporting a more European approach to space, or that it has hardly changed its position in this regard over the years, but that it started its pro-ESP campaign as early as 1979, if not earlier. Despite knowing that some Member States already had their own space programmes, and despite knowing that the Europeanisation of space beyond the ESA intergovernmental framework had hardly any chance of being accepted in 1979, the EP tried to motivate the decision-makers to intensify their collective efforts on space. As things turned out, its efforts were not wasted, even if they were not the only reason why an ESP came into being (Sigalas 2012).

The motive behind the EP's campaign may have been the enhancement of the EC's competences and eventually of its own powers, but it would have been impossible to state this openly. So how did the EP back its argumentation that Europe should play a greater role in space? First and foremost, by maintaining that such a move was beneficial. In its first Resolution in 1979, the EP suggested that the EC would benefit greatly from investing in space. This was not only in terms of scientific and industrial research, but also in terms of advances in the telecommunications, air traffic control, shipping control, earth observation, materials science, biology and medical research sectors. In short, investing in space would result in many benefits for diverse sectors.

Almost as early, from 1981 onwards, the EP started complaining in its Resolutions that the space programmes of the Europeans were either not as ambitious as those of their international competitors, at the time mainly the USA and the USSR, or that the competitors were far more advanced or committed to space. This situation, in the EP's view, was untenable and unacceptable: Europe had to catch up, Europe had to respond to a particular challenge and Europe had to solve a problem.

The potential benefits Europe could reap from space, on the one hand, and the challenges Europe would have to react to, on the other, were the two recurring and overarching themes that permeated all EP Resolutions, which I call super-frames. The kind of benefits to be drawn from space are simply too many to refer to in detail here. The breakdown of the super-frames to lower level frames will, however, give an idea about the nature of these benefits. Similarly, the kinds of challenges Europe has to tackle are many and diverse. I call them 'problem-solving' (or 'problem-reaction'), although they are not always problems in the strict sense. For example, the remark that there is an 'unprecedented crisis in the EU space industry' (Resolution P5_TA(2003)0217) would count as a problem-solving super-frame, but so would the position that the 'scope of the ESA programmes and of the Member States is limited' (1981 Resolution in OJ C 260/102), or that 'a significant portion of uses of space research are controlled by the market' (Resolution A5–0119/2000). What all three of these EP arguments have in common is that they suggest change is necessary. They imply that the 'unprecedented crisis' should be dealt with, that the 'scope of the ESA and Member State space programmes' should grow, and that the market should either not control such a 'significant portion of the space research uses' or that because of this it is necessary to act accordingly. If the benefits deriving from space are a positive incentive for action, problem-solving is the equivalent negative incentive.

Figure 4.2 presents the distribution of the benefit and problem-solving super-frames across time in absolute numbers. In addition to these two categories, I have included a residual category, which contains all the other justification arguments that could not be classified as either benefit or problem-solving. This suggests that the EP did not always justify intervention in terms of advantages to be gained. However, as Figure 4.2 shows, this is the exception. In most cases the EP calls for steps to be taken because of the advantages that are supposed to follow. What the figure also reveals is that the frequency of the benefits and problem-solving arguments has grown continuously over the years. With a few exceptions, the emphasis lies more on the benefits space can deliver (e.g. 'independence of Europe', OJ C 127/42) and less on the problem-solving (e.g. 'avoid unnecessary duplication of Member State activities', P6_TA(2008)0365) or problem-reaction ('private sector cannot finance all European infrastructure', A4–0384/1997) dimensions. Apparently, the EP does not want to overdo it with Europe's problems. It is perhaps more sensible to try to convince decision-makers by presenting space in a positive

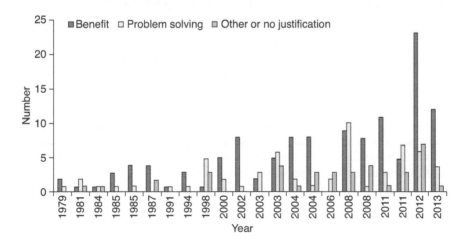

Figure 4.2 Space policy justification super-frames of the European Parliament (source: author's own calculations based on data about the Resolutions (see Table 4.2)).

light, rather than constantly reminding them how many problems Europe is facing. This is what the comparatively greater variation in the numbers of the problem-solving super-frames suggests, but there is probably another reason as well. Real problems, like the collapse of the public–private partnership for Galileo in 2007, appear at irregular intervals and this means more variation over time.

The growth of the benefit and problem-solving references over time shows that the EP has found this justification strategy increasingly relevant and useful. The mix of positive and negative incentives legitimating action on space matters is a powerful combination because it makes the justification discourse look more natural, which, in turn, makes it more convincing. However, Figure 4.2 also reflects what I noted earlier, namely, that as the years go by and the ESP takes shape, so do the details of the benefits to be gained or the problems to be solved. In other words, the increased use of the super-frames is only in part the outcome of a calculated choice. The rest is the natural consequence of policy growth.

What is not accidental is the choice to frame space in utilitarian terms. The EP prompts European institutions and politicians to act on space and develop a more comprehensive space policy not because it commands so, or because of some lofty ideals, but because there is something concrete to be gained (see also Chapter 5). This strategy allows the EP to evade the Eurosceptic critique that it is merely interested in advancing its own powers or those of the EU. By naming a number of advantages that the European industries, the citizens, the Member States or Europe as a whole have to gain, the EP is providing reasons why the current situation is sub-optimal and why Europe should move away from it. In other words, the utilitarian super-frames are a means of supporting, justifying and legitimating the development of Europe's space policy.

The EP justification frames for a European space policy

If Europe has only to gain from space, what has it got to gain exactly? The EP Resolutions provide not one answer, but many. They are of all sorts: general or particular, vague or precise, expected or surprising. Many of the answers have substantive similarities to each other, which allows them to be grouped into 19 categories or frames.

Space is usually deemed to be important in the EP Resolutions, either because of its many applications, or because of the importance of its applications. This is the reason why the two concepts can be grouped together. As I show in Figure 4.3, the importance of space and/or of its applications is the single most common justification frame. More than 14 per cent of all EP justification arguments high-light how important space is and how many uses it has for the industry, the scientific community and for ordinary citizens.

The second most common frame (8 per cent), and one that made its appearance very early on, is European independence (see also Chapter 1). Some of the typical arguments in this category are: Europe has to have independent access to space; Europe will become independent (usually from the USA); or space independence will have positive consequences for Europe. Equally important is the

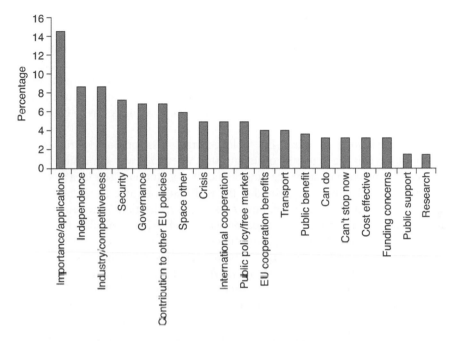

Figure 4.3 Space policy justification frames of the European Parliament (source: author's own calculations based on data about the Resolutions (see Table 4.2)).

frame labelled 'industry/competitiveness', which denotes industry-specific advantages or gains in economic competitiveness. This is often linked to another frame, namely, competition with non-European (again, mostly US) space industries, which are allegedly in a better or more competitive position than the European space industry.

One of the surprising frames found in the EP Resolutions on space is the connection between space and security. It is a surprising finding not because one does not expect a connection, but because it is made explicit. The dual-use (civilian and military) potential of space is the elephant in the room. Whereas it is clear that space infrastructure can and may be used for security (i.e. military) purposes, Galileo is normally advertised as a purely civilian infrastructure. Whatever the future may hold for the EU's flagship space programme, the EP does not shy away from the security relevance of space. As many as 7.21 per cent of all EP arguments justify the ESP in terms of security enhancement and there is even a whole EP Resolution dedicated to the topic of space and security (P6_TA(2008)0365).

As interesting and almost as frequent, each making 6.76 per cent of all arguments, are the two frames that come after 'security'. 'Governance' and 'contribution to other EU policies', as the labels suggest, deal with arguments referring to the institutional aspects of the ESP and with space contributing to the goals of other EU policies, respectively. With regard to governance, one would have expected the EP to try to ensure that the emerging ESP bodies and agencies are accountable to it. This means that one would have expected the EP to talk more about governance in its space Resolutions. Instead, the respective frame ranks fifth in terms of frequency. In my view, this should be interpreted as another example of the EP's strategic thinking. If baking comes before eating the cake, similarly consolidating the ESP comes before fighting for control of it. In other words, to decide who controls whom or what in the ESP, there has to be an ESP in the first place.

The contribution to other EU policies frame is particularly remarkable because it reveals the ingenuity of its inventor, whether it was someone from the EP, the Commission or someone outside the EU institutions. The idea is simple, yet brilliant. This frame suggests that the ESP did not come to replace, but to complement, and not to threaten, but to assist, the realisation of the other policy areas. Furthermore, it provides a reason why the EU has expanded its competences – not because the EU is power-hungry, but because space is the means to an end and the end has been agreed before. For instance, the EP argues that Galileo contributes to the 'realisation of the Europe 2020 strategy' (P7_TA(2011)0250) or 'satellite-based services contribute to the EU's Digital Agenda' (P7_TA(2013)0534). Finally, linking space with other EU policies makes the already long list of advantages significantly longer. On top of the usual benefits deriving from space, investing in space means investing in benefits deriving from other EU policies, the impact of which may be easier to appreciate. Hence space policy borrows legitimacy from the legitimacy of long-established EU policies that already have a track record of achievements.

International competition is a recurring theme in the EP Resolutions, covering 5.9 per cent of all space justification arguments. I name it 'space other' here, because it is nearly always linked to the notion of a non-European space power that poses a challenge or even a threat to Europe. It may be a powerful and long-established 'other' (the USA), a less powerful, but rapidly growing, 'other' (China, India), or even a no longer existing 'other' (the USSR before its dissolution). The presence of the 'other', as we know from psychology (Tajfel 1978), from nationalism studies (Neumann 1999) and also from European studies (Hörber 2012), plays an important role in the formation of the 'I' or the 'we'. There is little doubt that the EP's references to an 'other' in space affairs serves the purpose of giving shape to a European 'we', to a collective European space policy distinct from the individual, national level European policies.

The remaining legitimation frames each contain less than 5 per cent of all ESP justification arguments. They may be less frequent, but this does not mean that they lack in legitimation potential. The 'crisis' frame refers to problems specific to the European space sector, such as limited financing or delays in the Galileo project (for more on Galileo, see Chapter 13). Their dramatic presentation suggests that they are grave problems that ought to be dealt with immediately. 'International cooperation' addresses the space-related benefits that will be enjoyed by other countries (usually of the developing world), or arguments suggesting that a common space policy will bring about more international cooperation. 'Public policy/free market' is the frame where issues of privatisation and public investment are raised.

'Transport' and 'research' are fairly self-explanatory frames. They refer to the benefits space will bring to these sectors. 'Can do' and 'Can't stop now' are among the less frequent frames (3.15 per cent each), but they are noteworthy. The former contains the discussion that Europe should not be afraid to cooperate on space for it already has what is needed (the technological, institutional or legal capacity). The latter emphasises that the EU cannot stop now because any disruption in the space programmes would jeopardise what has been achieved so far. 'Cost-effective' and 'funding concerns' are conceptually related frames. In short, the first says that a common space policy will result in savings and the efficient use of resources by avoiding, for instance, the duplication of effort. Funding concerns, as the name suggests, deals with the need to find and guarantee sufficient funding for space. Lastly, a more recent and relatively minor frame, for the time being at least, connects space to the European citizens by arguing how important public support and understanding are for the development and success of the ESP.

Continuity and change in the ESP justification frames

The static analysis of the EP's justification frames for the ESP is indispensable in giving an overview of the importance of each frame in relative terms, but it can also be misleading if we forget that time is not a constant. A justification frame that was important in the early years of the ESP may not be as relevant 40 years later. Space was not mentioned in the EC Treaties in 1979 and the EP did not have the powers it enjoys today. Europe was different then and so was the rest of the world. It is therefore only natural that the legitimation frames will change over time. However, it is difficult to predict the details of change. For example, we lack a theory of change for the EP's legitimation frames. To offer but a sketch here, change is likely to depend both on extra-parliamentary factors (the political and economic environment at the national, European and global levels) and intra-parliamentary factors (the EP powers and composition, the portfolio breadth of the responsible EP committee, the rapporteurs' expertise). In addition, factors specific to the space sector (technological progress, the state of the space industry and market) and to the ESP (degree of maturity) are also likely to matter. Hence depending on if, in which way, and by how much these factors change, the EP is likely to vary its legitimation frames in a way that will allow them to continue resonating among the European decision-makers and the public.

Without a fully-fledged theory of change, it is not possible to explain exactly why the EP's justification frames changed, but exploring the Resolutions allows us to at least determine if and when any change took place. Table 4.2 summarises the distribution across time of the 12 more salient frames.

As expected, the space policy legitimation frames are not distributed evenly across time, albeit with some very interesting exceptions. Starting with the latter,

Table 4.2 Space policy justification frames of the European Parliament over time

Year	Importance/ application	Independence	Industrial competitiveness	Security	Governance	Contribution to other EU policies	Space (other)	Crisis	International cooperation	Public policy free market	Can't stop now	Can do
1979	–	1	–	–	–	–	–	–	–	–	–	1
1981	2	–	–	–	1	–	–	–	–	–	–	–
1984	–	–	1	–	–	–	1	–	1	–	–	–
1985	–	–	1	1	–	–	1	4	–	–	–	1
1985	2	1	2	–	1	–	–	–	–	–	–	–
1987	1	–	–	–	–	–	–	–	1	–	–	1
1991	1	–	–	–	–	–	1	–	–	–	–	–
1994	1	–	1	–	1	–	–	–	1	–	–	–
1998	1	–	–	1	–	–	2	–	1	4	–	–
2000	2	1	2	–	–	–	–	–	–	2	–	–
2002	3	–	1	–	–	–	–	–	1	–	1	–
2003	–	1	–	1	–	1	1	1	2	3	–	2
2003	1	1	–	–	–	3	2	1	1	–	–	–
2004	2	1	1	1	–	2	1	1	1	1	–	1
2004	1	–	–	1	–	–	–	1	1	–	–	–
2006	–	1	–	9	3	–	–	–	–	–	–	–
2008	–	–	–	1	2	2	2	1	–	–	2	–
2008	2	1	3	–	–	–	–	1	–	–	–	–
2011	2	2	1	–	2	1	1	1	–	–	1	–
2011	–	1	–	1	1	–	–	–	–	–	1	–
2012	8	5	2	1	2	4	–	1	2	–	2	1
2013	3	3	4	–	2	2	1	–	–	1	–	–
Total	32	19	19	16	15	15	13	11	11	11	7	7
No. of Resolutions frame appears in	15	12	11	8	9	7	10	8	9	5	5	6

Table 4.2 shows that the 'importance/applications' frame is not only the most frequent, but also the most persistent. The EP has relied on this reason almost uninterruptedly since 1981 to justify action on space matters. However, the oldest frame is European independence in space. Given that it is not a composite frame like 'importance/applications', it is fair to say that not relying on others to access and use space is the single most important argument for the birth and consolidation of the ESP. Whether it will continue to remain relevant after Galileo has granted Europe some independence in space remains to be seen.

Economic competitiveness and the space policy benefits for the European industry feature in as many as half of the Resolutions, implying that this frame is also very important. Interestingly, it first appeared in the 1984 Resolution – that is, only two years before the signing of the Single European Act. Space competition is the fourth legitimation frame that is characterised more by continuity rather than change. Like the other three, it is also one of the oldest frames. In the earlier Resolutions, Europe was overshadowed by the Cold War superpowers (the USA and the USSR); in the later Resolutions, the USSR has been replaced by the rising powers of China, Russia and India.

Security emerged as a justification frame relatively late, in 1985. It then disappeared for as many as 15 years, but since 2000 the EP has not hesitated to stress the link between space and security. Governance-related issues appeared sporadically in the earlier years and regularly after 2006. It seems that after the public–private partnership collapsed and Galileo became EU property, the 'governance' frame became regular. As I explained earlier, the EP appears to have prioritised the consolidation of the ESP before it started claiming any scrutiny and control powers.

The 'contribution to other EU policies' frame was singled out earlier as especially interesting and Table 4.2 confirms that this was for a good reason. Before 2003 this particular frame simply did not exist. It was first adopted around the time that the EU introduced the Europe 2020 strategy, which pledged to make Europe the most competitive economy in the world. The calculation behind this move is obvious. If the EU is committed to enhancing a knowledge economy or competitiveness, growth and jobs, then arguing that space can play an important role in this respect improves the chances that attention and resources will be directed to a common space policy.

'Can do' and 'can't stop now' have also been deployed strategically. The former was used primarily in the earlier period of space policy, whereas the latter was primarily used in subsequent years. It clearly makes more sense to persuade someone they can succeed in something if they have just started trying. Similarly, one way to encourage them to keep on trying is by reminding them how much they have already invested and how much they will lose if they stop now. Every time the EU is bound to start something new in the ESP, we should expect the 'can do' frame to reappear. Equally, the more the EU commits itself in space, the more salient the 'can't stop now' discourse is bound to be.

Conclusion

Space is expensive and politically sensitive. Developing a space policy demands the commitment of substantial sums of public money and reassurances that the resulting infrastructure will be used in a legitimate way. As a result, political leaders need to justify their policy choices in order to legitimate them. In the case of the ESP, the legitimation challenge is even greater. There have to be compelling reasons for pooling national sovereignty and resources to formally share competences on space policy with the EU, a supranational entity.

The EP, as I demonstrated in this chapter, took the legitimation challenge seriously. Being the only directly elected EU institution, it is compelled to defend its views publicly. It is not obliged, however, to favour Europe's involvement in space, but it does. The analysis of nearly all the EP Resolutions from 1979 to date revealed that the EP has been continuously supportive of a common policy in space from early on. Whatever its motives, over the years the EP has developed a powerful justification arsenal that has been deployed strategically to promote the development of Europe's own space policy.

To motivate Europe's decision-makers and stakeholders to take space more seriously, the EP has adopted a utility-focused legitimation strategy. The EP presents space as an advantageous endeavour in essentially all its Resolutions. On the one hand, it maintains that there are many and significant benefits to be gained if the Member States act collectively; on the other, collective action is deemed necessary because there are a number of specific problems or general challenges that can no longer wait.

Among the most important reasons legitimating investment in a common space policy is the need for Europe to shake off its dependence on other space powers such as the USA. Another very common legitimation frame is the importance of space as a result of its many applications in different sectors. Similarly, space has been consistently portrayed as a policy area that will deliver many benefits to Europe's industry and economy in terms of enhanced competitiveness.

This analysis of EP Resolutions has revealed that the EP is not only an ardent supporter of the ESP, but that it is also a strategic actor. Over the years the EP has introduced new frames, or re-introduced older ones, to maximise the appeal and resonance of its argumentation. Thus at the time the Single European Act was signed and the notion of a single European market became dominant, the EP started emphasising the competitiveness potential of space. Similarly, years later when the Europe 2020 strategy was launched, the EP started arguing that investment in space would contribute to more growth and jobs in Europe.

Strategic variation of the legitimation frames ensures that the ESP has increased the chances that it will continue to grow in the future. However, this does not depend only on the EP, even though its co-decision powers imply that no progress can take place without its consent. It would be very interesting to know how the other two key players, the Commission and the Council, have justified policy action on space. Such a comparative analysis would put the foundations for studying the causal relationship between legitimation frames and the take-off of EU policy. It is likely that the deepening of European integration will depend on the deployment of certain legitimation frames, but this is something for future research to prove.

Note

1 The views expressed in this chapter are of the author alone and not of the European Commission or any other organisation the author is affiliated to.

References

Chong, D. and Druckman, J. (2007a) 'Framing theory'. *Annual Review of Political Science*, 10: 103–126.

Chong, D. and Druckman, J. (2007b) 'A theory of framing and opinion formation in competitive elite environments'. *Journal of Communication*, 57(1): 99–118.

Druckman, J. (2001) 'On the limits of framing effects: who can frame?' *Journal of Politics*, 63(4): 1041–1066.

Entman, R. (1993) 'Framing: towards clarification of a fractured paradigm'. *Journal of Communication*, 43(4): 51–58.

European Union (2010) 'Consolidated versions of the treaty on the European Union and the treaty on the functioning of the European Union'. *Official Journal of the EU, C83*, 1–388.

Grimmer, J. and Stewart, B. (2013) 'Text as data: the promise and pitfalls of automatic content analysis methods for political texts'. *Political Analysis*, 21(3): 267–297.

Hörber, T. (2012) 'Refinements of antagonism in discourse theory for European studies'. *Journal of European Integration History*, 18(2): 207–220.

Krippendorff, K. (2004) *Content Analysis: An Introduction to its Methodology*. Thousand Oaks, CA: Sage.

Littoz-Monnet, A. (2014) 'The role of independent regulators in policy making: venue-shopping and framing strategies in the EU regulation of old wives cures'. *European Journal of Political Research*, 53(1): 1–17.

Nillsson, M., Nilsson, L. and Ericsson, K. (2009) 'The rise and fall of GO trading in European renewable energy policy: the role of advocacy and policy framing'. *Energy Policy*, 37: 4454–4462.

Neumann, I. (1999) *Uses of the Other: 'The East' in European Identity Formation*. Manchester: Manchester University Press.

Nugent, N. (2010) *The Government and Politics of the European Union*. Basingstoke: Palgrave Macmillan.

Sigalas, E. (2012) 'The role of the European Parliament in the development of EU space policy'. *Space Policy*, 28(2): 110–117.

Tajfel, H. (1978) *Differentiation between Social Groups*. London: Academic Press.

Valkenburg, P., Semetko, H. and De Vreese, C. (1999) 'The effects of news frames on readers' thoughts and recall'. *Communication Research*, 26(5): 550–569.

5 Europe's new wilderness

The Council's frames on space policy

Harald Köpping Athanasopoulos

'Where there is no vision, the people perish.'[1]

(Proverbs 29:18)

Introduction

The role of Europe in space has undergone a dramatic transformation since the turn of the twenty-first century. The EU is in the process of building its very own constellation of navigation satellites (Galileo), as well as constructing a system for earth observation (Copernicus, formerly known as GMES).[2] The European Space Agency (ESA), which has been implementing the space policy of its Member States since 1975, has seen its proportional expenditure for satellite navigation and earth observation skyrocket between 2000 and the present day, reflecting the increasingly important role of space applications. This chapter examines the role of the Council of the EU in the newly developing EU space policy. The Council represents the most intergovernmental EU institution, while also holding the largest inventory of decision-making powers. The ESA, in turn, is similarly intergovernmental and its vast body of experience in space makes it the obvious agent of an EU space policy. Hence the Space Councils were born out of the EU–ESA Framework Agreement of 2003, bringing together EU and ESA Member States at the ministerial level (Council of the European Union 2003a). Nevertheless, as will be demonstrated, the initial enthusiasm about the Space Councils has withered away and their significance continues to decline.

The purpose of this chapter is not to praise the role of the Council in EU space policy, but to highlight that its role can actually be found in a more subtle realm, which has traditionally not been associated with the Council. The Council has facilitated the transformation of the European space programme by selecting the frames in which it is presented. Any policy issue is inherently multidimensional because different actors are usually involved and because different interests are at stake: 'different policy actors' thus 'focus their attention on different aspects of the policy as they seek to build support for their positions' (Baumgartner and Mahoney 2008: 436). Although the European Commission needs to convince the Council (and the European Parliament) of the need for a particular policy, we might expect the Council's frames to have to be in line with the viewpoints of national governments and hence also with public opinion. As one of the two key legislators, the Council thus has the role of selecting the dominant frame from the variety of frames presented by the Commission.

In this early phase of the EU space policy, two competing sets of frames can be identified: (1) the value-rational frames of inspiration and exploration and (2) the instrumental rational frames of security, prosperity and independence. Although value-rational frames were initially drawn upon to explain the need for space-based activities, the new, application-driven approach is dominated by the prosperity and security frames. The changing frames reflect the Weberian conflict between instrumental rationality and value-rationality and fall in line with the postmodernist argument of Lyotard (1979) that knowledge is losing its 'use-value'. Lyotard puts forward the idea that knowledge is no longer gathered for its own sake, but always for some, usually economic, purpose. This argument will be elucidated in the following section.

After contextually placing the frames of the EU space policy into the Weberian dichotomy of value-rationality and instrumental rationality, this chapter will outline a brief history of the Council's involvement in an evolving European space programme. It will embark on an analysis of the Commission's Green Paper on Space Policy (Council of the European Union 2003b), which is a key document providing a number of competing frames for the justification of the EU's involvement in space. In an analysis of five crucial Council documents on space policy between 2003 and 2011, it will be demonstrated how the dominance of instrumental rational frames came about.

Competing frames and the transformation of knowledge

In the introduction to his book *Postmodern Condition*, Jean-François Lyotard 'define[s] *postmodern* as incredulity towards metanarratives' (Lyotard 1979: xxiv). While the age of modernity was shaped by an extreme appeal to myths and grand narratives, such as communism or scientism, postmodernism assumes that these approaches have failed. Harvey (1990: 43) sees modern hierarchy as being replaced by postmodern anarchy, and modern purpose replaced by post-modern play. The signs of postmodernity can be seen in nearly every aspect of society. In architecture, the formalism of the modern era has been abandoned in favour of a return to symbolism and a less rigid and structured style. This development can also be witnessed in the 'de-ideologification' of political culture in Europe and North America. Boggs (2000: 26) laments that 'familiar ideological traditions ... have lost their point of contact with rapidly changing global conditions', asking what ' "socialism" meant to the ruling parties of France, Spain and Greece' as the ideologies seem to still exist in their names, but not in their practice. An appeal to ideology no longer secures votes and society is increasingly elite-driven and technocratic, a state which Crouch (2004) refers to as 'post-democracy'. Crucially, in the postmodernist era, the role of knowledge has undergone a dramatic transformation. Lyotard suggests that:

> the relationship between suppliers and users of knowledge to the knowledge
> they supply and use is now tending ... to assume the form already taken by
> the relationship of commodity producers and consumers to the commodities
> they produce and consume – that is, the form of value.
>
> (Lyotard 1979: 4)

He recognises that if 'knowledge ceases to be an end in itself, it loses its "use-value"' (Lyotard 1979: 5).

Weber (1947) categorised purposeful social action into four categories, two of which are of particular interest in this context. Instrumental rational action (*zweckrationales Handeln*) refers to action that is meant to achieve concrete practical results. Parsons (1961: 970) states that 'one type or level of rationality concerns maximising results at minimum cost.... It includes economic rationality and rationality in the pursuit and use of political power'. The instrumental rational pursuit of space-based activities would therefore have to carry some material benefit. Value-rational action (*wertrationales Handeln*), on the other hand, is motivated by a 'conscious belief ... in the unconditional *intrinsic* value of a particular object, independent of success' (Weber 1947: 12). Parsons (1961: 970) refers to value-rationality as 'literal acceptance of the implications of an ultimate value-commitment'. A space programme is thus value-rational if the activity of space exploration itself is valuable, if a human presence in space is intrinsically desirable. Weber moreover discusses 'affectual action', which is caused by emotions, and traditional action, which is caused by habit (Weber 1947). The dichotomy between instrumental rationality and value-rationality will be maintained, as this categorisation helps us to understand the initial competition between frames and the eventual dominance of instrumental rational frames for space policy.

Lyotard's observation of knowledge losing its use-value implies that the pursuit of knowledge, and thus the pursuit of space-based activities, must become increasingly driven by the potential economic benefits as well as by security considerations. It can thus be expected that instrumental rational frames are dominant. Historically, discoveries in astronomy have had few practical consequences, but were nevertheless deemed to be inherently valuable. If the nature of knowledge is indeed undergoing a transformation towards losing this inherent value, we would anticipate space-based activities to be justified as a means towards some valuable (presumably economic or security-related) end. The chapter will now examine to what extent this hypothesis holds true and, as the role of human agency in this process is emphasised, it will analyse whether the Council has played a significant role in this change.

The Council's involvement with space

To assess the extent to which the Council has contributed to the reorientation of Europe's space policy, it is necessary to first examine how this reorientation has occurred practically. As the ESA is the agent that is implementing the EU space policy, its annual budget is a good starting point for visualising these changes. Within a period of four years, navigation and earth observation have become the main foci of the ESA (European Space Agency 2007, 2012), a development which ESA official Charlotte Mathieu describes as an 'evolution towards applications' (personal communication, 15 June 2011). This correlates with the EU involvement with the ESA and former Commission President Barroso describing Copernicus (then GMES) and Galileo as the EU's 'flagship programmes' in space (Barroso 2009). Budget allocations for the exploration of the solar system as well as for human spaceflight have stagnated, which becomes especially apparent when taking an annual average inflation rate of 2 per cent into account. Although the ESA's overall budget has vastly increased over the past eight years, it appears as though policy-makers are interested in space applications, rather than in space exploration.

A close analysis of Galileo and Copernicus further reveals a tendency towards the militarisation of space. 'Militarisation' is here understood to refer to the use of space for military purposes, as opposed to 'weaponisation', which refers to the placing of weapons systems in orbit around the earth (Mutschler and Venet 2012). The European Commission claims that Galileo's Public Regulated Service (PRS) aims to deliver a more precise signal than the standard Galileo signal, which will be available free to the general public (European Space Agency 2010). A survey by the Galileo Supervisory Authority reveals that 'half of all users of the Galileo encrypted signal [PRS] will be military customers' (GPS World 2008). Despite the mantra of Galileo being a 'civilian system under civilian control', it is clear that it is being built in full awareness that military customers will be one of the major sources of revenue (Mörth 2000: 178). Nevertheless, any satellite navigation system can potentially be used by the military and the question of whether PRS is going to be used to guide missiles remains unanswered. Copernicus, the second 'flagship programme', has similar military applications. A Copernicus service that was developed in the G-MOSAIC project, 'will provide intelligence to the EU and its Member States before and after a crisis occurs' (European Commission 2009). One of the project's reference users is 'EU military staff', further highlighting Copernicus' military potential. Sheehan (2009) poetically refers to this development as 'profaning the path to the sacred' because the ESA Convention explicitly states that the exploration of space is 'for peaceful purposes' only (European Space Agency 1998).

Before directly addressing the frames on space policy, it will be shown that the Council's significance in these developments is not in its role as a legislator, but lies elsewhere. For this purpose, a brief history of the formal involvement of the Council with EU space policy is needed. Many points of departure could be chosen to narrate a brief history of the Council's involvement with space policy, but the single most meaningful event occurred in 2003 with the signing of the ESA–EU Framework Agreement (Council of the European Union 2003a). As already mentioned, the intergovernmental nature of the Council and the ESA appears to pre-designate the two institutions to close cooperation.

In reality, it was, of course, the European Commission that negotiated the Framework Agreement 'on behalf of the Community' (Council of the European Union 2003a), with the Council merely signing the agreement, an important albeit functional role. The agreement emphasises the 'peaceful use of space', as well as the potential for 'economic growth', with the basic intention behind the agreement being to avoid 'any unnecessary duplication of effort' (Council of the European Union 2003a). The Agreement mentions eight specific fields of cooperation, which cover nearly all elements of the ESA's activities, including, of course, navigation and earth observation, but also 'human space flight and micro-gravity' (Council of the European Union 2003a). The only major field of activity that was omitted was the robotic exploration of space. An intense debate surrounded the possible trajectories for future cooperation between the ESA and EU (e.g. Hörber 2009; Sheehan 2009), with suggestions reaching from case-by-case cooperation to the incorporation of the ESA into the EU institutions. The reality of EU–ESA collaboration has been remarkably close to the vision out-lined in the Framework Agreement. The latter foresees the 'management by the ESA of European Community space-related activities', as well as the 'carrying out of activities which are coordinated, implemented and funded by both parties' (Council of the European Union 2003a). Both strategies have been put into prac-tice, most prominently in the Galileo and Copernicus programmes.

Although the actual signing of the Agreement is a noteworthy aspect of the Council's involvement with the EU–ESA Framework Agreement, Article 8 of the document potentially has real significance as it establishes the Space Coun-cils, which are defined as 'regular joint and concomitant meetings of the Council of the European Union and the Council of ESA at ministerial level' (Council of the European Union 2003a). The institutional similarities between the two bodies come into play here and the eight Space Councils that have been facilitated since the signing of the Agreement provide useful markers in an analysis of the Coun-cil's role in European space policy (Table 5.1). Since the first Space Council in 2004, the event has usually been held during the Competitiveness Council ses-sions and sometimes shortly before the regular ESA Council, where ideas brought up at the Space Council can potentially be implemented.

The Space Councils have experienced a rather parabolic development, in which the peak of their importance was reached around 2009 and then a steady decline can be demonstrated ever since. The primary focus of the first two meet-ings appears to have been the preparation of the subsequent Space Council. A major issue of concern was the 'identification of priorities of the ESP, including estimation of costs' (Council of the European Union 2004). The third Space Council, in turn, focused on Copernicus, having played an important role in turning the project into one of the EU's 'flagship programmes' in space. Although the first three Space Councils developed 'orientations', the fourth Space Council in 2007 was the first to produce a Resolution (Council of the European Union 2007). The Resolution was in line with the Lisbon Agenda, highlighting 'the emerging European knowledge society' (Council of the Euro-pean Union 2007), yet it also stresses the security dimensions of Copernicus and

Table 5.1 Outcomes of the Space Councils

	Date	Outcome
First Space Council	25 November 2004	Preparing programme of next Space Council in spring 2005 (Council of the European Union 2004)
Second Space Council	7 June 2005	Preparation of future European Space Programme; defining roles and responsibilities of actors involved, particularly the EU and ESA (Council of the European Union 2005a)
Third Space Council	28 November 2005	Orientations on Copernicus, which was to provide information for policy-makers on security and the environment (Council of the European Union 2005b)
Fourth Space Council	22 May 2007	First resolution on European space policy, focus on growth and creation of a knowledge-based economy; focus on security dimensions of Copernicus and Galileo; dual-use potential of space; dialogue with European Defence Agency (Council of the European Union 2007)
Fifth Space Council	26 September 2008	Resolution focuses on Galileo and Copernicus as the flagship programmes; identification of four priority areas for the European Space Programme: • Tackling climate change • Space as a means to implement Lisbon Treaty • Contribution of space to Europe's security • Space exploration (Council of the European Union 2008)
Sixth Space Council	15 June 2009	Resolution puts even greater emphasis on the security dimension of space, mentioning MUSIS and SSA (Council of the European Union 2009)
Seventh Space Council	25 November 2010	Resolution emphasises Galileo and Copernicus and abandons references to MUSIS; section on cooperation with Africa (Council of the European Union 2010)
Eighth Space Council	6 December 2011	Resolution has a very strong focus on security; detailed examination of SSA; highlights the importance of completing Galileo and Copernicus (Council of the European Union 2011a)

Galileo. The Council saw the need for a 'structured dialogue' between the EU and the 'European Defence Agency for optimizing synergies' and recognises the dual-use potential of many space-based systems (Council of the European Union 2007).

The following 2008 Resolution on space (Council of the European Union 2008) placed much greater emphasis on Galileo and Copernicus, with nearly half of the document devoted to the two programmes. In the 2009 Space Council (Council of the European Union 2009), the security dimension of space-based activities was pushed further into focus. References were made to MUSIS, a failed European military surveillance satellite system, as well as to SSA (Space Situational Awareness), a project aimed at monitoring near-earth objects, including foreign satellites. Although the former application has never been implemented, both systems have clear military purposes, highlighting the role of the Space Council in pushing the dual-use potential of space onto the agenda. Although the preparatory stage of SSA has been completed, it remains to be seen whether the programme as a whole will be put into practice. The 2010 Space Council meeting abandoned references to MUSIS, yet underlined that the 'GMES programme will allow Europe to deal more effectively with global security issues' (Council of the European Union 2010). Furthermore, the document contains a section on cooperation with Africa, emphasising the potential benefits of the European Geostationary Navigation Overlay Service (EGNOS) for the continent, a system designed to maximise the operability of the US global positioning system for European and, potentially, African users.

The most recent Council Resolution on space, which followed the Eighth Space Council in 2011, is shorter than previous Resolutions and focused almost entirely on security (Council of the European Union 2011a). Although it contains hardly any new material, it clearly reveals the priorities for the Council in space: exploiting its security potential. The Resolution contains a more detailed examination of SSA and follows the tradition of all previous Resolutions in highlighting that the completion of Copernicus and Galileo is paramount for Europe. Nevertheless, the repetitiveness of the Resolutions, particularly after 2009, shows the declining importance of the Space Council meetings, which also became significantly shorter. Although the first Space Council was scheduled to take two hours, the latest meeting required a mere 65 minutes (Council of the European Union 2011b). Indeed, the space section on the Council's internet presence contains links to all documents produced by the Space Councils, apart from the 2011 Resolution. This not only accentuates the decreasing significance of the meetings, but it puts the transparency of the Space Council sessions into question.

The Council's frames on space

The main argument advanced in this chapter is that the Council's role in European space policy was not in shaping and proposing new programmes, but in determining the frames used to justify European space endeavours. For this purpose, the following six documents will be examined:

1   The European Commission's Green Paper on European Space Policy
     (Council of the European Union 2003b). Although this is a Commission
     document, it is included in this analysis because it is the first extensive EU
     document on space policy published in January 2003. Furthermore, it sets
     up archetypical representations of different competing frames for space
     policy, from which the Council would be able to select the dominant frame.
2   The EU–ESA Framework Agreement (Council of the European Union
     2003a). Published in October 2003, this document draws on the Green Paper
     and represents a major hallmark in the history of EU space policy.
3   The Resolution on the European Space Policy that following the fourth
     Space Council (Council of the European Union 2007). This is the first major
     document that clearly outlines the Council's vision of the EU's role in
     space.
4   The Resolution following the fifth Space Council (Council of the European
     Union 2008). This document is a very clear formulation of the Council's
     ideas and it is important as it establishes Galileo and Copernicus as the cor-
     nerstones of the EU space programme.
5   The European space strategy (Council of the European Union 2011c). This
     is one of the most important documents on the EU's role in space and it is
     very useful for analysing the competing frames as a draft version is avail-
     able (Council of the European Union 2011d).
6   The Resolution following the eighth meeting of the Space Council (Council
     of the European Union 2011a). This document is the most recent Council
     Resolution on space.

Looking at these documents chronologically allows us to draw conclusions about
the impact of the Council on defining appropriate frames (Table 5.2).

The Commission's Green Paper on European Space Policy sets itself the task
of initiating 'a debate on the medium- and long-term future use of space for the
benefit of Europe' (Council of the European Union 2003b: 6). The document
was developed after a request by the European Parliament and relied on a 'series
of five consultation workshops' organised by 'the EC/ESA Joint Task Force'
(Council of the European Union 2003c). These workshops were meant to contribute the views of different interest groups and there was thus one workshop
for the industrial community and one for the scientific community. Contrary to
expectations, the industrial workshop was indeed very supportive of the ESA's
scientific activities and encouraged further funding for the International Space
Station. The consultation report on the Green Paper does not inform the reader
about the outcomes of the open web consultation, which gave members of the
public the opportunity to voice their opinions. The 2003 White Paper, however,
mentions the conclusions of the web consultation, singling out a call for the
'support of the exploration of the solar system' as well as the 'need for a long-
term vision including human spaceflight' (Council of the European Union
2003d). In fact, the instrumental rational view that space should be used primarily for economic growth and job creation is not particularly visible in any of

Table 5.2 The Council's space frames

Document	Date	Frame
Green Paper on Space Policy	24 January 2003	Document sets up two value-rational frames (exploration and inspiration) and three instrumental rational frames (prosperity, security and independence); the dominant frames will emerge from this selection
EU–ESA Framework Agreement	7 October 2003	Emphasis on prosperity to justify EU engagement with ESA and in space in general
Fourth Space Council Resolution	25 May 2007	References to prosperity, security and independence, with particular emphasis on prosperity; some references to inspiration and exploration
Fifth Space Council Resolution	29 September 2008	Emphasis on prosperity, in combination with security; independence frame is present and, to a lesser extent, also the exploration frame
European Space Strategy	31 May 2011	Different versions show how the inspiration frame had been replaced by the prosperity frame
Eighth Space Council Resolution	6 December 2011	Strong emphasis on the prosperity frame; important role of the security frame in justifying space endeavours

the workshops, although it is mentioned in some. The workshop for the scientific community is noteworthy because 'a plea was made to stop the trend to significantly reduce the funds for European space science research' (Council of the European Union 2003c).

The Green Paper, which was the final product of these workshops, begins its introduction with the following paragraph:

> 'Last Frontier...', 'discovery of the Universe and its origins...', '...life on other planets...', '...first footsteps on the Moon...', '...space heroes...'. Space represents to humanity an infinite, timeless source of dreams and striking reality.
>
> (Council of the European Union 2003b: 6)

The Green Paper seeks to inspire its readers, reminding them of the deep effect that the view of the stars has had on humanity. It essentially states that the EU cannot ignore this 'timeless source of dreams'. referring to 'space heroes' and the 'last frontier' – space is a challenge that humanity simply must engage with (Council of the European Union 2003b). Going to space is seen as valuable in itself and two value-rational frames are set up:

1 The inspiration frame puts forward that space is a source of dreams for humanity. A human presence in space is desirable simply because it inspires.
2 The exploration frame advocates that exploring, discovering and understanding the universe is part of human destiny.

The Green Paper nevertheless also sets up several other frames, which can be categorised as instrumental rational. The document states that:

> a strong European presence in particular key areas of space applications is indispensable both as a political asset and in order to enable the Union to maintain its strategic independence and contribute to its economic competitiveness.
>
> (Council of the European Union 2003b: 19)

This sets up three distinct instrumental rational frames for space policy, which became recurrent in the Council's publications on space:

1 The security frame argues that space is important for 'monitoring hazardous transport operations' or 'border surveillance' (Council of the European Union 2003b: 19).
2 The independence frame proposes that Europe has a strategic interest in independent access to space, which is why an EU space policy is important.
3 The prosperity frame argues that space is critical to the European economy and that it is therefore paramount for the EU to possess a space policy if it intends to remain highly competitive.

The document does highlight, however, that 'space is … a high-risk sector, of fragile economic viability' (Council of the European Union 2003b: 6). This statement was, of course, underlined by the failure of the public–private partnership that was initially meant to fund Galileo. Nevertheless, it establishes two competing sets of frames, allowing the Commission to choose which frame will become dominant.

The Framework Agreement between ESA and the EU has already given us a hint about the Council's preferred choice of frames. Although the Agreement makes no attempt to justify space-based activities, it mentions in the first paragraph of Article 1 that the European Space Policy ought to 'link demand for services and applications using space systems in support of the Community policies with the supply of space systems in infrastructure necessary to meet that demand' (Council of the European Union 2003a). This should be seen as a reference to the Lisbon Agenda and is hence exemplary of the instrumental rational prosperity frame. It clarifies the Council's perspective on why the EU needs a space policy: to meet the demands of its economy. Space is clearly a means to an end rather than an end in itself and, while human space flight is mentioned as a specific area of cooperation, it is unclear to what extent this can be accommodated under a supply-and-demand rationale.

The 2007 Resolution on the European space policy that followed the fourth Space Council provides more detailed insights into the Council's ideas for a European space policy. Although a section with the title *Vision for Europe* evokes expectations for value-rational justifications, the Council looks at space as a 'strategic asset', which can 'contribute to the independence, security and prosperity of Europe' (Council of the European Union 2007). It thus contains explicit references to all three instrumental rational frames. Reference is made to the Lisbon Agenda, identifying the potential of space to generate 'growth and employment' (Council of the European Union 2007). As this is the first justification for going to space that is mentioned in the Resolution, it is evidence of the Council pushing the instrumental rational prosperity frame into the spotlight. The Council recognises the 'inspirational ability of space activities', hence referring to the inspiration frame, but wants to use it for 'attracting young people into science and engineering' (Council of the European Union 2007). Space exploration has historically had this inspirational ability precisely because it is seen by many to be an end in itself; the Council, however, wants to use this ability to implement the Lisbon Agenda, turning it into a means to an end. Nevertheless, the Resolution also contains a paragraph referring to the exploration frame, 'recognising that the exploration of space contributes to answer far-reaching questions on the origin and evolution of life in the Universe' (Council of the European Union 2007), an aim which is not economically or militarily motivated. Hence the two sets of frames are still present and competing.

The next Council Resolution of 2008, Taking Forward the European Space Policy, took on a very similar discourse, with a further emphasis on instrumental rational frames. The document's introduction contains an explicit reference to the independence frame, reaffirming the 'importance for Europe to maintain an autonomous access to space' (Council of the European Union 2008). The 'vision for Europe is space' that is outlined in the document sees the primary objective of the European space policy in strengthening 'Europe as a world-class space leader responding to the needs of European policies and objectives, in terms of applications, services and related infrastructures' (Council of the European Union 2008) This is an important reference to the prosperity frame. It includes a separate section in the Lisbon Agenda, containing more such references, using particularly instrumental rational terms such as 'economic opportunities', 'growth', or 'economic exploitation' (Council of the European Union 2008). The security frame is also present, although in combination with the prosperity frame. The Resolution states that 'space assets have become indispensable for our economy' and that 'their security must thus be assured' (Council of the European Union 2008). However, to a lesser extent, the value-rational exploration frame is referred to as well and a 'human expedition to Mars' is mentioned (Council of the European Union 2008). Space exploration is nevertheless seen as a 'global endeavour' that ought to be conducted 'within a worldwide programme' (Council of the European Union 2008), hence justifying the moderate European funding for robotic exploration. The Resolution shows how the dominance of the prosperity frame is being established out of the initial variety of frames.

In 2011, the Council worked on a document outlining the European space strategy and the differences between draft versions emphasise competition between frames as well as the increasing dominance of the prosperity frame. The short section on space exploration is particularly striking as there seems to have been significant internal debate about the precise wordings of each point. Although a draft version emphasises the importance of space 'for the benefit of all mankind' (Council of the European Union 2011d), the same point in the final version calls for an examination of 'possible options for involvement in space exploration setting out a cost benefit analysis' (Council of the European Union 2011c). Furthermore, the section on space exploration was made significantly shorter. The two versions highlight the contrast between the instrumental rational and value-rational frames for going to space. The draft version is a reference to the inspiration frame; it is reminiscent of the plaque left on the Moon by the crew of Apollo 11 and the heroism that is inevitably connected to the Moon landings. Space exploration is regarded as an inherent good, requiring no further justification. The final version, on the other hand, appeals to notions of prosperity, measuring the value of space exploration in economic terms. This rationale becomes more evident as we read the remainder of the document, which refers to a 'strict economy of resource' and 'user-driven applications', emphasising that 'space activities and applications are vital to our society's growth and sustainable development' (Council of the European Union 2011c). The two versions not only show that the Council is adopting an increasingly instrumental rational discourse on space, favouring the prosperity frame, but they also demonstrate that particular formulations are chosen consciously. Competing forms of rationality seem to exist within the Council, with instrumental rationality seemingly being given priority; as the title of the document states, a European space strategy has to 'benefit' the citizens (Council of the European Union 2011c).

Following the release of the 'space strategy', the Space Council met once more, creating a further Resolution. Reference is made to the Europe 2020 strategy in the opening section and there is an emphasis on the 'role which space systems play to provide ... practical tools for ... the implementation of European policies' (Council of the European Union 2011a). The Council's rhetoric is dominated by the instrumental rational prosperity frame, arguing that space exploration will enable 'economic expansion and new business opportunities' (Council of the European Union 2011a). However, apart from the prosperity frame, the security frame is a similarly important theme in the Resolution, which is abundant in references to 'European and non-European citizens' safety and security requirements' (Council of the European Union 2011a). The section on space exploration, on the other hand, contains few novelties, being partially assembled using phrases from previous documents. By 2011, we can therefore conclude that the instrumental rational frames and the prosperity frame in particular have become thoroughly dominant in the Council's discourse on EU space policy.

Conclusions: the Council as a frame selector

The Lisbon Treaty has fundamentally altered the European supranational decision-making process as the European Parliament has been granted co-decision powers. Nevertheless, the Council arguably remains the more influential institution in the new bicameral politics of the EU (Hagemann and Høyland 2010). This is visibly demonstrated by the current economic crisis, where the Council is central to discussions surrounding the future of the Eurozone, as opposed to the absence of the Parliament and even the Commission. It was our goal for this chapter to investigate the role of the Council in the shaping of EU space policy. Space policy is an example of the interplay between the Commission as the frame-creator and the Council as the frame selector; this is arguably a microcosm of inter-institutional relations. Whether the Parliament takes on a similar role remains the subject of another potential study. Nevertheless, as a result of the difficulty involved in attempting to access the minutes of all the relevant meetings, one of the weaknesses of the argument presented here is that it cannot be sufficiently shown to what extent the Council's discourse has practically led to a reorientation of the ESA programme.

Lyotard's argument was that knowledge is becoming ever more commodified, which is a development that this analysis of the Council's frames for space policy vividly validates in two ways. Space applications providing services and generating economic or security gains have greatly gained in importance. Copernicus and Galileo are not only the EU's 'flagship programmes' in space, but they also constitute the largest parts of the budget of the ESA, which has undergone a radical transformation. While the European space exploration programme remains ambitious, and while much was achieved using modest financial resources, it has hardly benefited from the massive investments into ESA during the past decade. It appears to be easier to justify space-based projects that 'benefit the citizens of Europe' economically than to argue that we should develop more powerful rockets for missions to Mars. Although this instrumental rational reasoning stands on uneven ground, as Galileo seems to fail in delivering the promised superiority to the US global positioning system in terms of signal quality, it has nevertheless generated a revitalised European interest in space. On the other hand, the fact that space exploration is justified by making reference to job creation and economic expansion further underlines the cogency of Lyotard's point.

The meeting of the ESA Council at Ministerial Level in November 2012 ran under the slogan 'Space for competitiveness and growth'. Funding for a larger version of the European heavy-lift rocket Ariane 5 was secured and the ESA Member States agreed to fund several probes to Mars in collaboration with Russia. The Lunar Lander programme was scrapped. Europe must make sure that the constant references to cost–benefit analysis and international cooperation do not take on a ritualistic character, preventing serious investment in space exploration (e.g. the Aurora programme). The idea of the human exploration and colonisation of the solar system can provide a vision in a postmodern society where knowledge has become a mere means to an end. Using space as a catalyst to foster human curiosity is essential if the human spirit of exploration is to be the guiding ideal for future European integration.

Note

1 Scripture quotation from The Authorized (King James) Version of the Bible. Rights in The Authorized Version in the United Kingdom are vested in the Crown. Reproduced by permission of the Crown's patentee, Cambridge University Press.
2 In this chapter, the European Earth observation programme that was formerly known as GMES will be referred to as Copernicus. It should be noted, however, that 'GMES' and 'Copernicus' are synonymous.

References

Barroso, J.M. (2009) *The Ambitions of Europe in Space*. SPEECH/09/476. Brussels: European Commission.

Baumgartner, F.R. and Mahoney, C. (2008) 'The two faces of framing: individual-level framing and collective issue definition in the European Union'. *European Union Politics*, 9(3): 435–449.

Boggs, C. (2000) *The End of Politics*. New York, NY: Guilford Press.

Council of the European Union (2003a) *Council Decision on the Signing of the Framework Agreement between the European Community and the European Space Agency*. Document Number ST 12858 2003 INIT1, 7 October 2003.

Council of the European Union (2003b) *Green Paper on European Space Policy*. Document Number ST 5707 2003 INIT, 24 January 2003.

Council of the European Union (2003c) *Green Paper on European Space Policy: Report on the Consultation Process*. Document Number ST 14886 2003 ADD 1, 17 November 2003.

Council of the European Union (2003d) *White Paper: Space: a New European Frontier for an Expanding Union*. Document Number ST 14886 2003 INIT, 17 November 2003.

Council of the European Union (2004) *Report on Proceedings in the Council's Other Configurations*. Document Number ST 15631 2004 INIT, 9 December 2004.

Council of the European Union (2005a) *Situation Report on Proceedings on the Council's other Configurations*. Document Number ST 9734 2005 INIT, 9 June 2005.

Council of the European Union (2005b) *Report on Proceedings in the Council's Other Configurations*. Document Number ST 15213 2005 INIT, 7 December 2005.

Council of the European Union (2007) *Outcome of the Proceedings of the Council (Competitiveness) on 21–22 May 2007 – Resolution on the European Space Policy*. Document Number ST 10037 2007 INIT, 25 May 2007.

Council of the European Union (2008) *Council Resolution: Taking Forward the European Space Policy*. Document Number ST 13569 2008 INIT, 29 September 2008.

Council of the European Union (2009) *Council Resolution on the Contribution of Space to Innovation and Competitiveness in the Context of the European Economic Recovery Plan, and Further Steps*. Document Number ST 10500 2009 INIT, 29 May 2009.

Council of the European Union (2010) *Council Resolution: Global Challenges: Taking Full Benefit of European Space Systems*. Document Number ST 16864 2010 INIT, 26 November 2010.

Council of the European Union (2011a) *Council Resolution: Orientations Concerning Added Value and Benefits of Space for the Security of European Citizens*. Document Number ST 18232 2011 INIT, 6 December 2011.

Council of the European Union (2011b) *Preparation of the Competitiveness Council of 5 and 6 December 2011*. Document Number ST 17238 2011 REV 1, 25 November 2011.

Council of the European Union (2011c) *Council Conclusions on Towards a Space Strategy for the European Union that Benefits its Citizens*. Document Number ST 10901 2011 INIT, 31 May 2011.

Council of the European Union (2011d) *Draft Council Conclusion on Towards a Space Strategy for the European Union that Benefits its Citizens*. Document Number ST 8815 2011 INIT, 15 April 2011.

Crouch, C. (2004) *Post-democracy*. Malden, MA: Polity Press.

European Commission (2009) *Space Research: Let's Embrace Space*. Brussels: EC Publications Office.

European Space Agency (1998) *ESA Convention* [online]. Available from: www.esa.int/convention/. Last accessed 24 May 2011.

European Space Agency (2007) *Annual Report 2007* [online]. Available from: www.esa.int/esapub/annuals/annual07/ESA_AR2007b.pdf. Last accessed 23 November 2012.

European Space Agency (2010) *Galileo Specifications* [online]. Available from: www.esa.int/esaNA/SEMTHVXEM4E_galileo_0.html. Last accessed 19 June 2011.

European Space Agency (2012) *Funding* [online]. Available from: www.esa.int/SPECIALS/About_ESA/SEMNQ4FVL2F_0.html. Last accessed 23 November 2012.

GPS World (2008) Galileo gloves come off: military after all. *GPS World,* issue 10, August.

Hagemann, S. and Høyland, B. (2010) 'Bicameral politics in the European Union'. *Journal of Common Market Studies*, 48(4): 811–833.

Harvey, D. (1990) *The Condition of Postmodernity: an Enquiry into the Origins of Cultural Change*. Oxford: Blackwell.

Hörber, T. (2009) 'ESA + EU: ideology or pragmatic task sharing?' *Space Policy*, 25(4): 206–208.

Lyotard, J. (1979) *The Postmodern Condition*. Manchester: Manchester University Press.

Mörth, U. (2000) Competing frames in the European Commission – the case of the defence industry/equipment issue. *Journal of European Public Policy*, 7(2): 173–189.

Mutschler, M. and Venet, C. (2012) 'The EU as an emerging actor in space security?' *Space Policy*, 27(2): 118–124.

Parsons, T. (1961) *Theories of Society: Foundations of Modern Sociological Thought*. New York: Free Press.

Sheehan, M. (2009) 'Profaning the path to the sacred'. In: N. Bormann and M. Sheehan (eds), *Securing Outer Space*. London, New York: Routledge.

Weber, M. (1947) *Grundriss der Sozialökonomik: Wirtschaft und Gesellschaft* (own translation). Tübingen: Mohr.

6 Role of the European Commission in framing European space policy

Lucia Marta and Paul Stephenson

Introduction

In the Lisbon Treaty, which entered into force on 1 December 2009, space was for the first time explicitly mentioned as a direct and shared competence of the EU (Article 4.3 and 189). Although no institution is mentioned in terms of providing leadership in this domain, or being solely responsible for steering developments, it is in practice the European Commission, the EU's policy entrepreneur, which is largely responsible for shaping space policy. This chapter traces the way in which the Commission has framed (and reframed) the issue of EU space programmes and services in recent years. It investigates how the agenda-setter has 'talked about' space policy, with a particular focus on Galileo and Copernicus, examining how the Commission's own institutional discourse – as revealed in its communications throughout the programme's 'definition' and 'implementation' phases – has evolved over time by way of 'frame sets'. In so doing, it shows how the EU's agenda-setting executive has chosen to present the issue of space involvement as politically, technologically and economically desirable for the EU, and how it has sought to persuade decision-makers, European taxpayers and even itself of its cross-policy relevance and potential benefits.

The Commission as a space policy actor

Right of initiative, competences and windows of opportunity

The Commission enjoys the right of initiative under the ordinary legislative procedure. It has mainly issued communications on space to the Council and the European Parliament (EP) from a range of different issue perspectives. Communications have concerned, for instance, a European space policy (Commission 2011a, 2012a), a space industrial policy (Commission 2013a), joint European programmes and endeavours for economic growth (Commission 2013a), such as Galileo (see Chapter 2; Court of Auditors 2009; Stephenson 2012; Commission 2013c), as well as strategies for sustainable development using space (Commission 2011c, 2012c) and general governance of space activities in Europe and its relation with other actors, such as the European Space Agency (ESA) (Commission 2012b). Most communications have ended up, after debates and modifications by the co-legislators, as being effective – although non-binding – legislative tools. The Commission has also proposed regulations, such as for Copernicus (Commission 2013c), and decisions to the Council and the EP, as with the future Space Surveillance and Tracking service (Commission 2013c).

A few proposals have not made it very far – for instance, the communication proposing the establishment of an intergovernmental fund to finance the operation of the Copernicus programme from 2014 to 2020 (Commission 2012). Member States and the EP were opposed to this on the grounds of the supposed detrimental effects it would have on programme development and as a sign of EU disengagement from a flagship initiative (European Parliament 2012). The European Economic and Social Committee adopted a similar position (European Economic and Social Committee 2013). Eventually, the Commission agreed to re-insert Copernicus financing into the EU budget, although the amount was reduced.

In the late 1990s, space programmes were exclusively in the hands of Member States and international organisations, such as the ESA and the European Organisation for the Exploitation of Meteorological Satellites (EUMET-SAT). The EU did not yet have competences in the sector, so to overcome over this hurdle the Commission used its competences in the areas of research and technological development (R&TD) to 'insert' space-related projects into the EU's R&TD agenda. It defined the research objectives to which space activity could potentially contribute. An historical policy footprint remains today, demonstrated by the structure and wording of the Lisbon Treaty: space-related articles are inserted in title XIX, *Research and Technological Development and Space*.

In addition to R&TD policy, two other key policies were used at the time to frame the need for space in terms of its contribution to their objectives. Thanks to Copernicus – the earth observation programme – the EU would be able to obtain up-to-date and reliable data relevant to environment protection and the monitoring of climate change. The same can be said for Galileo, the satellite navigation and positioning system (see Chapter 15), which has been mostly oriented to providing better traffic management and transport safety, although the benefits will spill over to other policies.

The Commission has enjoyed several windows of opportunity in which to introduce space-related initiatives to its agenda. The Kyoto Protocol (signed December 1997 and effective February 2005) was important as far as environmental protection was concerned. The Balkan conflicts (1991–1999) and the so-called 'revolution in military affairs' placed pressure on the EU and showed its incapacity to carry out military intervention in an autonomous way through space-based and networking capabilities. European industrial welfare was also at stake following a wave of privatisation running through the aerospace and defence industries in the USA. This provoked a political reaction, with the restructuring of European firms and the recognition of the need to coordinate

European efforts in terms of industrial consolidation and the financing of R&TD activities and programmes. These external factors allowed the Commission to make a case for the EU to develop its own advanced technologies that would enable it to compete globally, while exploiting its own space systems for security purposes. Thus not only was there an environmental or transport safety frame, but also frames of 'market opportunity', 'technological and industrial development' and 'non-dependence in security and defence' matters, in coherence with its R&D initial competence.

Commission resources

In terms of the resources at its disposal, these can be differentiated between 'legal–political' and 'financial'. First, it has boasted the power of initiative in the legislative process (since shared with the European Council), but also legal competences of policy execution in a range of policy domains, including research, environment, transport, as well as fisheries and agriculture, to name but a few. These are policy areas whose management largely benefits from space applications. The Commission could thus 'pitch' space policy as delivering important tools to improve or ensure technologically advanced policy implementation. It could create a more favourable regulatory environment for these policies by framing and promoting the use of technologies by end-users, both at the institutional level (national and European, such as civil protection and law enforcement authorities) and at commercial and 'citizen' level (such as insurance and the scientific community). By engaging in a process of identifying and organising trans-European end-user communities, the Commission contributed to the commercial uptake and use of common space-based services, which provide high-tech market opportunities for European firms. Indeed, those firms needed to ensure a follow up on their R&D activities and to secure a return on their R&D co-investments.

Second, through the preparation and proposal of the European Multiannual Financial Framework (MFF), the Commission has been able to better plan budgetary expenditure across a range of common policy areas and programmes such as Copernicus and Galileo, as well as engaging in agenda-setting for the framework programmes for R&TD. In so doing, it has ensured co-financing for European firms for their research activities (see Chapter 10). The agreed 2014–2020 MFF foresees more than €4 billion for the implementation and operation of the Copernicus programme and about €7 billion for the Galileo programme. The MFF earmarks about €80 billion for the Horizon 2020 programme (current prices), in charge of co-financing R&TD activities, of which about €17 billion (22 per cent) is for the section dedicated to 'industrial leadership'. Space features within this section, with about €1.5 billion of co-financing over seven years, which includes R&TD for the two main flagship programmes, but also research in the domains of science, technologies for launcher and the protection of space infrastructures (European Parliament and Council 2013).

The Commission has encountered financial difficulties in ensuring appropriate funding for these programmes. For Galileo, the initial plan was to finance it through a public–private partnership – that is, with the financial participation of the private sector. The public–private partnership failed and in 2007 the EU decided to seek full financing from the public budget (see Chapter 13). Galileo consists of infrastructures that belong entirely to the EU and therefore the EU remains its owner and main sponsor (together with the financial participation of the ESA). With respect to Copernicus, in 2009 the proposal for the 2014–2020 MFF clearly highlighted the Commission's difficulties in ensuring the necessary financing for large-scale projects (Commission 2011b), including Copernicus, which, after tough negotiations with the budgetary authorities, was reinserted into the EU budget. In spite of this, the Commission has given warning signals to the private and public partners concerning the huge financial effort required by space programmes and the need to ensure that over-run costs do not put them at risk.

In terms of human resources, the Commission has limited expertise. By its very nature, it tends to be run mostly by policy makers without technique expertise. Although these desk officers are engaged in elements of policy formulation, they largely rely on external scientific and technical expertise, industrial lobbies, as well as other international organisations (especially the ESA). The Commission staff do not necessarily possess the specific technical insight needed to develop state-of-the-art space projects or, subsequently, to manage them. Instead, the Commission seeks to shape and steer the agenda, define high-level objectives, identify policy priorities, flesh out the wording of new proposals and, more generally, provide impetus and momentum to the development of common European space activities. The formulation of frame sets also helps the Commission staff to better appropriate themselves with space programmes.

Directorate-General enterprise keeps the lead

Many Commission Directorate-Generals (DGs) are involved in space policy, given that most of them will be end-users of Copernicus and Galileo services – for example, the DGs for Competition, Environment, Agriculture and Fisheries, and the Transport and Maritime DGs. In terms of leadership and policy-making in the space field, however, the DG for Enterprise and Industries (DG ENTR) has been the main actor to date. In the last Commission up to November 2014, two main Directorates within DG ENTR were involved: one dealing with Copernicus and Space Policy (Directorate for Aerospace, Maritime, Security and Defence industries) and one with Galileo (Directorate for EU satellite navigation programmes) (Marta 2013a). Historically, Galileo has been managed by DG MOVE – that is, DG Mobility and Transport (formerly DG TREN – DG Transport and/or DG Transport and Energy – in its various guises): Antonio Tajani, former Director-General of this DG (May 2008–February 2010), took ownership of Galileo (one of the 30 trans-European network priority projects) and carried the portfolio with him when he left for DG ENTR.

Copernicus and Galileo are managed at a high level and with the particular political standpoint taken by DG ENTR. Both programmes have almost completed their R&TD phases and are entering their operational phase. That DG

ENTR is in charge of EU space programmes, at least politically speaking, high-lights the joint 'industry' and 'economy' frame of the two programmes: they are presented as essential tools for industrial and economic development and growth. Such a joint frame remains coherent with the R&D dimension, as firms are the key R&D players. This joint frame is continuously used in all institutional documents and presentations, as well as in communication tools used to promote the awareness of space activities among citizens. Examples include websites, press releases, larger publications (books, leaflets, even comic books) and ad hoc events such as the European Space EXPO, a free space-related exhibition visiting European capitals to 'show citizens how European space policy and space-based technologies benefit our everyday lives on earth and also of course, their importance for the European economy and job creation' (see DG ENTR website).

DG ENTR consults and coordinates with a number of other DGs whose policies will ultimately rely on space services. Special mention needs to be made of the Joint Research Centre (Ispra, Petten, Karlsruhe) in charge of scientific research, including some space-related activities. Its competences can be exploited at the implementation and operational levels, as is the case for the Emergency Management Service of Copernicus, created to support the management of emergency situations in Europe – such as fires, floods, earthquakes and industrial accidents. The Emergency Management Service is coordinated both by the Commission and the Emergency Response Centre, with the Joint Research Centre offering technical support.

The latest Commission has been operational from November 2014: while space was supposed, initially, to be managed by a new DG called Transport and Space to be headed by Maroš Šefčovič (from Slovakia), the space portfolio has once again been inserted into the DG for Internal Market, Industry, Entrepreneurship and SMEs (DG GROWTH), headed by Elżbieta Bieńkowska (from Poland). In his mission letter to the Commissioner, Juncker asks her to:

> [E]stablishing a coherent and stable regulatory framework for the service and manufacturing of space applications in Europe and exploiting the internal market and job-creating potential of space. This will include setting the conditions for the development of markets for space applications and services including the exploitation of space data, data from scientific missions and commercial applications of space data.
>
> (Juncker 2014)

The sudden changes that occurred in the structure of the Commission were due mainly to the need to re-balance nationalities and portfolios after the EP vote against a proposed Commissioner. Therefore the final structure is not the result of political logic or will. Space would have probably gained momentum if an ad hoc DG was dedicated to this field (e.g. DG for Transport and Space). Despite the importance attached to space in its discourses, this domain was added in a 'last minute move' on top of an already very full area of responsibility (Juncker

2014). It is too soon to observe the dominant frames operated by the new Commission, but it seems that the 'industrial and economic' frame is still the main one.

The Commission vis-à-vis other actors

The Council

The Commission is engaged in space policy alongside the EU's two co-legislators, the Council and the EP. In the EP, space is dealt with by a number of committees (see Chapter 4) and in the Council by the working groups and Space Councils (see Chapter 5). In its 'competitiveness' configuration, the Council discusses Copernicus, while the Transport, Telecommunications and Energy configuration of the Council deals with the Galileo programme (Marta 2013a). While the Commission may 'explore' Member States' intentions before drafting a communication or a proposal, in drafting those texts it uses frame sets and advances interests that have a communitarian purpose. The Commission engages in extensive dialogue with the Council and its Member States to ensure political equilibrium, as without the support of the national capitals no action would be possible.

The European External Action Service

The relationship between the recently created European External Action Service (EEAS) and the Commission is still evolving, both from a political and legal point of view. Despite the existence of a 'bridge' represented by the High Representative of the Union for Foreign Affairs and Security Policy/Vice-President of the European Commission, currently Federica Mogherini, both institutions need to fine-tune their relationship. This is also the case at the operational level. For instance, as the Space Surveillance and Tracking (SST) proposal (Commission 2013c) is worded, the two 'should define the coordination mechanisms needed to address matters related to the security of the SST support programme'. Moreover, as far as Copernicus' security service is concerned, if a key role is to be attributed to the EU Satellite Centre (SatCen), delegation agreements and legal tools will need to be developed between the Commission and SatCen (which receives operational direction from the High Representative). Moreover, in the new Commission structure, Juncker has entrusted a number of well-defined priority projects to Vice-Presidents, including the High Representative, acting on his behalf and implementing new powers. This may also entail an evolution in the relationship between the Commission and EEAS in space matters.

Finally, the EEAS is currently leading negotiations at the international level for the adoption of a Code of Conduct on Outer Space Activities. Although the Commission does not lead the initiative, which is of a political and diplomatic nature, it is engaged in complementary R&D activities. From a technical point of view, the EC contributes to the general purpose of the Code, which is increasing security and safety in space. An intense relation in space affairs between the EC and the EEAS would ensure that the Commission benefits from the extensive support and coordination structures of the EU's external affairs corps, historically anchored in the former second pillar, and crucial for space policy (NASA 2012).

Other EU agencies and bodies

Some Copernicus services benefit from a decentralised and delegated management structure, supported by existing European agencies and organisations possessing ad hoc expertise, such as the European Environment Agency and FRONTEX (the European Agency for the Management of Operational Cooperation at the External Borders of the Members States of the European Union). A key actor at EU level is SatCen in Torrejón, Spain, founded in 1992 by the Western European Union (WEU) – a defensive alliance composed of ten Member States founded in 1948 and modified in 1954 – and incorporated into the EU as an agency in 2002. It is in charge of supporting the Common Foreign and Security Policy and European Security and Defence Policy through space-based intelligence and is inserted in the EEAS structure under the political supervision of the Political and Security Committee of the Council, which issues guidance to the High Representative.

These agencies provide specific expertise to ensure the full exploitation and efficient management of space programmes. An ad hoc agency, the GNSS Supervisory Authority, was set up in 2004 to manage Galileo services, replacing the Galileo Joint Undertaking. Based in Prague, it is supported by the Galileo Security Monitoring Centre, based in Saint Germain-en-Laye outside Paris. The GNSS Supervisory Authority has an Administrative Board consisting of Member State representatives, a representative of the Commission and the EP, with the ESA and the High Representative as observers. The Commission is also assisted by a number of bodies within the EU institutions involving officials and national representatives with specific expertise, such as in the GMES (now Copernicus) Committee and its 'security board'. End-users and experts appointed by Member States meet in the User Forum to assist the Commission in the definition of user requirements.

European Space Agency and National Space Agencies

Beyond the EU institutions, the most important body with which the Commission has had relations is the ESA. Established in 1975 in Paris, the ESA – which is not part of the EU – has its own rules, procedures and approach to space activities. It has 22 Member States (20 EU countries, plus Norway and Switzerland) and holds regular Ministerial Councils (see Chapter 3). Canada also sits on the Council and takes part in some projects under a Cooperation Agreement. Today, the global European governance of space activities is largely the concern of the

EU, the ESA and the National Space Agencies of the Member States. Indeed, the ESA was active for 25 years before the Commission and national space agencies had existed long before this. Thus EU activities have been built up by drawing on existing scientific knowledge, experience and technological and industrial capabilities. Current coordination mechanisms and relations among these actors are set out in a general Framework Agreement between the EU and ESA (signed in 2003 and entered into force in May 2004).

The ESA supports the Commission in its pursuit of policy objectives through its technical, management and development skills. In 2013, the EU was the largest contributor to the ESA budget, to the sum of more than €900 million, far more than other international organisations (notably EUMETSAT). Generally speaking, a certain work-sharing exists among the main actors in accordance with their capabilities and competences. However, the relationship between the EU and ESA is not considered satisfactory, mainly by the EU, whereas ESA seeks to evolve to better serve European space interests in a new context (ESA 2012; Commission 2012b). The Commission has advocated that 'a closer relationship with the ESA would enable a further development of divisions of tasks' (Commission 2012b: 1). The Commission invited the Council and the European Parliament to provide feedback on its suggestions concerning the relationship between the EU and ESA and the long-term goal of rapprochement of ESA towards the EU framework. On such a basis the Commission could provide a cost–benefit analysis of the possible options (Commission 2012b: 5).

At times work and responsibility sharing have been or remain unclear between the two, both at the decision-making level for a common European space policy and at the management level. In fact, despite the fact that almost the same countries participate in both the EU and the ESA, the two maintain their own identities, approaches, interests and goals. In its communication, the EC lists the '[S]tructural obstacles in the current EU/ESA relations', including:

2.1. Mismatch of financial rules [...]
2.2. Membership asymmetry [...]
2.3. Asymmetry in security and defence matters [...]
2.4. Absence of mechanisms for policy coordination [...]
2.5. Missing political accountability for ESA [...]

(Commission 2012b)

As one academic puts it, there are 'two captains on the European spaceship' (von der Dunk 2003). Drawbacks arising from this situation have been analysed in the literature (Marta 2013b), but also put forward by the two institutions themselves (ESA 2012; Commission 2012b). Discussions are currently underway on this issue. During the ESA Ministerial Council in December 2014, ESA Member States declared the wish not to change the intergovernmental nature and approach of the Agency (ESA 2014b).

At national level France, the UK, Germany and Italy, as well as Belgium, Spain, Norway and Sweden, have significant national space programmes and policies. The Commission deals with representatives from the respective national space agencies. It has always being keen on exploiting existing national capabilities, as for Copernicus, the space component of which partly consists of existing national assets. The same applies to the proposal for an SST service, based on existing national capabilities that 'feed' the development and operation of a European service. Exploiting national assets responds mainly to the necessity of not duplicating costs and, arguably, respects subsidiarity. However, it requires compromise and coordination. Member States with an advanced technical capacity in space tend to retain control of their assets, of the information derived, as well as the technological know-how, mainly to protect their investments, security, commercial interests and national identity. Generally, Member States and national space agencies are careful to manage their relations with the Commission (as far as framework programmes for R&D funds are concerned) and with the ESA (as far as ESA-managed funds are concerned), given their individual political, economic and industrial interests.

The private sector

European industries provide technical expertise and technological development in the space market. The Commission has long dealt with private firms operating as part of a 'space lobby'. These firms, with experience in hardware construction, aerospace, metals and information technology, are keen for European space policy to include ambitious multi-billion euro programmes that will bring tenders and, potentially, contracts to supply new infrastructure. These firms include the large transnational aerospace and defence companies – such as Airbus Defence & Space (formerly EADS), Thales Alenia Space, Finmeccanica, GMV and Indra – but also formerly relatively small companies that are growing thanks to important EU space contracts, such as Orbitale Hochtechnologie Bremen (OHB), in charge of the Galileo satellites. Moreover, hundreds of small and medium-sized enterprises – with unique and extremely technical and refined know-how – are involved as suppliers. These firms are organised within a common association, the Aerospace and Defence Industries Association of Europe (ASD) based in Brussels, with a specific group dedicated to space called Eurospace. The European Satellite Operators' Association, of which companies such as Inmarsat, Eutelsat and SES are part, gathers together industries involved in the delivery of telecommunication services.

Trade associations ensure coordination among their European members for a common lobbying effort in the institutions. For instance, ASD is one of the few, if not the only, actor to have published a statement that warns of the effects that the Transatlantic Trade and Investment Partnership initiative, currently under negotiation, may have on European aerospace industries. It has called for the Commission to consider it carefully and to involve the ASD in its discussions. Dialogue between space-related industries and the Commission over specific EU space-funded programmes normally begin well before a contract for production is awarded. European industrial groups participate as leaders or partners in

R&DT projects while the service or infrastructure is being developed. Even beforehand, firms may participate in defining an R&D programme, as the Commission may not have the necessary technical skills to know which technologies and capabilities exist or need to be developed to respond effectively to common policy objectives. Private interests and the Commission act as partners, ensuring the 'policy problem' is met with a technological solution and, in so doing, contribute to the Commission's related goals (and, therein, frame sets) of 'economic growth', 'qualified jobs' and 'technological independence'.

Main space programmes and promoted frames

EU flagship space programmes: Copernicus

The EU, and the Commission in particular, uses the term 'flagship programme' to indicate its two most ambitious space programmes, Copernicus and Galileo. The recent proposal for a Regulation establishing the Copernicus programme aims to define and organise the operational and exploitation phase from 2014 to 2020 (Commission 2013c; European Parliament and Council 2014). It contains an explanatory memorandum in the introduction that highlights the arguments ('frames') used by the Commission to justify the programme. Accordingly, Copernicus 'aims at providing Europe with a continuous, independent and reliable access to observation data and information'. Copernicus was developed because 'public authorities need to access independent, reliable, timely information in order to take informed decisions' and because 'public authorities need information to support/better implement a wide range of policies'.

Commission communications claim that a wide range of users' communities will benefit from 'timely and reliable added-value information and forecasting to support for example, agriculture and fisheries, land use and urban planning, the fight against forest fires, disaster response, maritime transport or air pollution monitoring'. It will help secure 'economic stability' and 'smart and inclusive growth' given the commercial benefits that can be fostered by space (Commission 2013c). The proposal also contains arguments to justify Commission involvement in the programme. It asserts that 'continuing such observations is becoming critical, considering the increasing political pressure on public authorities to take informed decisions in the field of environment, security and climate change and the need to respect international agreements'. As for financing, it claims that 'responsibility for funding the exploitation and the renewal of space infrastructure developed with EU and intergovernmental funds cannot be optimally achieved by individual Member States because of the costs incurred':

> The provision of other services (e.g. emergency maps or thematic land monitoring maps of a more limited geographical scope) can be better achieved at EU level for two reasons. First, a more coherent and centralized management of input data, from space based or in-situ sensors, will allow for economies of scale. Secondly, a coordinated provision of earth monitoring

services at Member State level helps to avoid duplications and enhances the monitoring of the implementation of EU environmental legislation on the basis of transparent and objective criteria.... Moreover, action at European level will create economies of scale leading to a better value for public money. Action at EU level thus leads to a clear added value.

(Commission 2013c)

EU flagship space programmes: Galileo

Similar frames have been at play with Galileo. As for Copernicus, Galileo is intended as a technologically advanced programme, at the service of a large European community of end-users, both institutional and private. The 'independence' frame is put forward for institutions, especially in areas of potential application, such as security and defence. The 'market' frame is advanced to firms in terms of future commercial opportunities. Galileo is pitched as a programme for promoting economic growth, industrial development and the creation of jobs for qualified people. Those frame sets can be found in Commission documents over the different phases of the programme, from the development and definition phase (see Chapter 2) to current deployment:

The satellite navigation programmes ... are flagship projects of the Union. They aim to provide navigation services and will foster considerable developments in numerous sectors of activity, drive technological innovation and growth of competitiveness in the European economy and provide a source of job creation, commercial revenue and socioeconomic benefits. As such, they form part of the Europe 2020 strategy and policies for sustainable development. More specifically, the Galileo programme aims to establish Europe's global navigation satellite system (hereafter GNSS). It will lead to the provision of positioning, timing and navigation services to users worldwide for a wide range of applications, from transport, financial securities clearance, electricity provision, weather forecasting, to road tolling.

(Commission 2013d)

In 2011, the Commission used similar frames to introduce Galileo into its proposal concerning the exploitation and implementation of the European navigation systems:

Promoting this technology, which is pitched as being a powerful tool for helping the EU emerge from the financial and economic crisis, fits in perfectly with the Europe 2020 strategy and policies for sustainable development. [They] provide considerable opportunities for all fields of activity with many new jobs bound up with the expansion of markets, which have grown at an annual rate of 30 per cent over the past few years

(Commission 2011c)

In this context, the Commission has been working to develop 'an ecosystem of applications to optimise the use of services provided by the systems and to maximise the socio-economic benefits.... Moreover, small and medium-sized enterprises everywhere in Europe play an important role in the programmes' (Commission 2011c).

The Commission is investigating the potential commercial use of Galileo in selected areas, including smart phones, to ensure a return on its investment (Spacenews 2014). As far as EU added value is concerned, it has highlighted that:

> The systems established under the European satellite navigation programmes are infrastructures set up as trans-European networks of which the usage extends well beyond the national boundaries of the Member States. Furthermore, the services offered through these systems contribute, in particular, to the development of trans-European networks in the areas of transport, telecommunications and energy infrastructures. Satellite navigation systems cannot be set up by any single Member State as this would exceed its financial and technical capabilities. Therefore, it can be only achieved by action at EU level.
>
> (Commission, 2011c)

Space Surveillance and Tracking

More recently, the Commission has proposed establishing an SST support programme (Commission 2013b) aimed at providing a European service to institutional and private owners and operators who wish to protect their space infrastructures from space debris that may damage or destroy them. Therefore the 'security frame' was prevalent. However, the Commission has also used an 'economic frame' to push forward the SST initiative:

> The space surveillance and tracking proposal envisages the pooling of resources to protect our investments in space infrastructure from damage. Avoiding space collisions could save up to €210 million per year and remove a serious risk to the delivery of economic gains expected from the EU's space programmes. I welcome the approval of the Parliament for the Commission's SST proposal and hope for its swift adoption by the Council.
>
> (Commission 2014)

The Commission has underlined that 'progress on two flagship European programmes, Galileo and Copernicus ... has also raised awareness of the need to protect EU space infrastructure' (Commission 2013b). It has used the 'independence frame' to stress the need for the EU to reduce dependence from the USA. By the USA, 'sharing' space surveillance and tracking data, services aim to be fed by, French and German SSA assets and capabilities. 'The proposed European SST services accommodate an essential objective of the space industrial policy of the EU, namely to achieve European technological non-dependence in critical domains, and to maintain independent access to space' (Commission 2013b). The Commission justifies its role in this sector by way of the EU's new direct competence in the field of space (Lisbon Treaty) and with the will of the Member States: 'The need for EU action in the domain has been supported by Member States in several Council Resolutions and Conclusions' (Commission 2013b):

> The European SST service is to be led by the EU and not by the European Space Agency (ESA).... The underlying reason for this emerged in numerous discussions: the European SST service has a security dimension (it allows gathering intelligence on States' civil and military space infrastructure and operations) which the EU, unlike ESA, has the competence and is equipped to deal with. The TFEU grants the EU the competence to coordinate the exploitation of space systems and has also the competence and the mechanisms in place to deal with the security dimension of such a service; Member States consider that ESA should support the EU in this endeavour (and it is doing so through its SSA preparatory programme) but, as an R&D organisation, does not have the competence and the mechanisms necessary to set up and run a European SST service on its own.
>
> (Commission 2013b)

Despite such sensitivity, the proposal makes clear that 'The provision of SST services should serve only civilian purposes. Purely military requirements should not be addressed by this Decision' (Commission 2013b).

The Commission has promoted its role in SST in a different way to the flagship programmes. It approach is arguably less ambitious, aware that existing SST assets (such as radar and telescopes) belong to Member States and ESA. Previously, it advocated taking ownership, being engaged in management and taking a political lead. Conscious of the high military sensitivity of some of those assets and data derived (radars in particular, can be part of missiles' early warning systems) and deliverable services, the Commission has advocated the EU playing a more limited role:

> Member States ask the EU to define the governance and data policy for a European SST service, to play an active role in the setting up of the service, and to make best use of existing sensors and expertise. Member States are also explicit as to how security concerns should be taken into account: SST sensors need to remain under national control.... The Member States retain ownership and control over their assets and remain responsible for their operations, maintenance and renewal.
>
> (Commission 2013b)

The EU provides support through grants (including lump sums). The beneficiaries are participating Member States contributing with national assets to the European SST system, as well as the SatCen, where it cooperates with participating

Member States to establish and operate the SST service function. The overall indicative EU contribution to implement the support programme is €70 million for the period 2014–2020 at current prices (European Parliament and Council 2014). Horizon 2020 foresees also some research and development budget for a specific objective called 'protection of European assets in and from space' to which €6.5 million is allocated for 2014–2015 research and development activities (see H2020 website).

Conclusion

The role of the European Commission (and, in particular, DG ENTR) in the development of a European space policy has been key, not least in terms of agenda-setting. While the ESA has provided management expertise and scientific know-how and European firms have provided technical capabilities, the Commission has pushed for and shaped a European space policy using multiple frames in its communication. These frames are varied and complementary, depending on the programme involved, the phase of progress and the political window of opportunity. The Commission initially framed space as contributing to the achievement of EU policy objectives in research and development, climate change, environment, transport safety and the security of its citizens. These frames sought to give political legitimacy to the notion of EU activity in space, especially vis-à-vis the Council and the EP. By focusing on the tangible benefits that space programmes bring to all citizens across the continent, it sought to legitimate EU action to taxpayers.

Subsequent frames concerning market development and the provision of services to governmental and commercial end users helped make a case for the economic benefits to be derived from exploiting space infrastructures. Economic growth, the creation of jobs for qualified people and industrial competitiveness justified EU financial involvement. Moreover, by using a frame of technological development and independence, the Commission supported the identity creation of the EU as a technologically advanced society. The Commission developed these frames as the result of intensive dialogue with external actors, without whose support it could not hope to develop or finance a common space policy.

More recently, as the operational phase of the flagship programmes has approached, there has been a need to meet 'pragmatic' challenges. Thus the Commission has talked about the implementation and operational requirements of space programmes. It has spoken in much more practical terms about technical and financial elements, management, organisation and political governance as it continues to seek a balance of power between all the space actors involved. Nonetheless, it has been keen to maintain political support for projects, recalling older frames, in the face of delays and excessive costs and given the current economic crisis. As one Commission official put it at a recent conference, even at this stage of maturity, the EU budget for space cannot be taken for granted (Dr P. Weissenberg, Deputy Director-General, Directorate of Space, Security and GMES and in charge of the Galileo programme, personal intervention).

The new research and development frame (H2020) for the next seven years foresees about €1.5 billion for space projects across several areas of activity, including exploration and access to space, areas which until now have not resulted in concrete EU-led programmes. Coupled with the ongoing debate over the general governance of space activities in Europe and the ability of the Commission to propose itself as an actor with changing roles, it is logical to assume that the ambitions of the EU – and the Commission – in space will continue to grow and by no means be limited to either the current programmes or existing frame sets.

References

Commission (2011a) *Communication Towards a Space Strategy for the European Union that Benefits its Citizens.* COM/2011/0152 final, 4 April 2011.

Commission (2011b) *Communication from the Commission to the European Parliament, the Council, the European Economic and Social Committee and the Committee of the Regions: A Budget for Europe 2020.* COM(2011) 500 final, 29 June 2011.

Commission (2011c) *Regulation of the European Parliament and of the Council on the Implementation and Exploitation of European Satellite Navigation Systems.* COM(2011) 814 final, 30 November 2011.

Commission (2012a) *Communication from the Commission to the Council and to the European Parliament: European Space Policy.* COM/2007/0212 final, 26 April 2012.

Commission (2012b) Communication from the Commission to the Council and to the European Parliament Establishing Appropriate Relations between the EU and the European Space Agency. COM/2012/0671 final, 14 November 2007. Available from: eur-lex.europa.eu/LexUriServ/LexUriServ.do?uri=COM:2012:0671:FIN:EN:PDF.   Last accessed 15 May 2015.

Commission (2012c) *Communication from the Commission to the European Parliament, the Council, the European Economic and Social Committee and the Committee of the Regions on the Establishment of an Intergovernmental Agreement for the Operations of the European Earth Monitoring Programme (GMES) from 2014 to 2020.* COM/2012/0218 final, 11 May 2012.

Commission (2013a) *Communication from the Commission to the European Parliament, the Council, the European Economic and Social Committee and the Committee of the Regions: EU Space Industrial Policy – Releasing the Potential for Economic Growth in the Space Sector.* COM/2013/0108 final, 28 February 2013.

Commission (2013b) *Proposal for a Decision of the European Parliament and the Council: Establishing a Space Surveillance and Tracking Support Programme.* COM(2013)0107 final, 28 February 2013.

Commission (2013c) *Proposal for a Regulation: Establishing the Copernicus Programme and Repealing Regulation (EU) No. 911/2010.* COM(2013)0312 final/2, 29 May 2013.

Commission (2013d) *Proposal for a Regulation of the European Parliament and of the Council Amending Regulation (EU) No. 912/2010 Setting up the European GNSS Agency.* COM(2013) 40 final, 6 February 2013.

European Commission (2014) *Avoiding Costly Space Crashes: European Parliament Approves Space Surveillance and Tracking Programme.* Press Release, 2 April 2014. Available from: www.europa.eu/rapid/press-release_IP-14-364_en.htm. Last accessed 15 May 2015.

Court of Auditors (2009) *Concerning the Management of the Galileo Programme's Development and Validation Phase Together with the Commission's Replies.* Special Report No. 7/2009, 26 June 2009.

Dunk, F. von der (2003) 'Towards one captain on the European spaceship – why the EU should join ESA'. *Space Policy* 19: 83–86.

European Space Agency (2012) *Political Declaration Towards the European Space Agency that Best Serves Europe, Naples, November 2012.* Luxembourg: Ministerial Council.

European Space Agency (2014) *Resolution on ESA Evolution.* Luxembourg: Ministerial Council.

European Economic and Social Committee (2013) *Opinion on the Proposal for a Regulation of the European Parliament and of the Council Establishing the Copernicus Programme and Repealing Regulation (EU) No 911/2010.* COM(2013) 312 final – 2013/0164 (COD), 16 October 2013.

European Parliament (2012) *Resolution of 16 February 2012 on the Future of GMES.* 2012/2509(RSP), 16 February 2012.

European Security Research Advisory Board (2006) *Meeting the Challenge: the European Security Research Agenda Report.* Luxembourg: Office for Official Publications of the European Communities.

European Parliament and Council (2013) *Regulation (EU) No. 1291/2013 of 11 December 2013 Establishing Horizon 2020 – the Framework Programme for Research and Innovation (2014–2020) and Repealing Decision No. 1982/2006/EC.* Available from: www.eur-lex.europa.eu. Last accessed 15 May 2015.

European Parliament and Council (2014) *Decision No. 541/2014/EU Establishing a Framework for Space Surveillance and Tracking Support.* Available from: www.eur-lex.europa.eu. Last accessed 15 May 2015.

Juncker, J.C. (2014) *Mission Letter to Elżbieta Bieńkowska, 1 November 2014.* Brussels: European Commission.

Marta, L. (2013a) *Europe: Spread (Not Lost) in Space.* Brief Issues No. 22. Paris: European Union Institute for Security Studies.

Marta, L. (2013b) 'National visions of European space governance: elements for a new institutional architecture'. *Space Policy,* 29(1): 20–27.

NASA (2012) *Presentation to the 49th Session of the Scientific and Technical Subcommittee – Committee on the Peaceful Uses of Outer Space – United Nations, 6–17 February 2012: USA Space Debris Environment, Operations, and Policy Updates.* Available from: www.unoosa.org/pdf/pres/stsc2012/tech-26E.pdf. Last accessed 15 May 2015.

Spacenews (2014) *Europe Weighs Galileo-compatibility Mandate for Smartphones.* Available from: www.spacenews.com/40214europe-weighs-galileo-compatibility-mandate-for-smartphones/#sthash.jE8RFBvw.dpuf. Last accessed 15 May 2015.

Stephenson, P.J. (2012) 'Talking space: the European Commission's changing frames in defining Galileo'. *Space Policy,* 28(2), 86–93.

7 Role of the Court of Justice of the European Union in the development of EU space policy

Antonella Forganni

Introduction

In the area of space law, the presence of 'two Europes in one space' (Madders and Thiebaut 1992: 117) is rather peculiar. The 'two Europes' are the European Space Agency (ESA), an intergovernmental organisation, and the European Community (later the EU), a supranational entity. The overlapping of the two organisations is even more interesting if we consider that the Member States of the ESA and the EU are not identical.

On the one hand, Article XVII of the ESA Convention provides that any dispute between Member States, or between Member States and the ESA, regarding the interpretation or application of the Convention, or the immunities and privileges, should be settled by the ESA Council or by arbitration. On the other hand, space policy is a shared competence of the EU, as stated by Article 4 of the Treaty on the Functioning of the EU (TFEU) and, as such, it is included within the jurisdiction of the Court of Justice of the European Union (CJEU).

This chapter considers the contribution of the CJEU in framing space policy. Although the CJEU is not a political institution, its interpretation of EU law can lead to the framing or reframing of EU policies, especially in a field such as space that is still underdeveloped. On the basis of the methodology described by Stephenson (2012: 88), deconstructing the process by which an issue is framed implies 'examining how problems and solutions are articulated regarding an issue'. A selected group of the CJEU's judgements, divided into sub-groups, is analysed in this chapter, which highlights how the Court has approached certain questions, how it has qualified the related legal issues and which answers have been provided.

Space policy includes several fields that are characterised by the presence of different interests, such as security, industry, science and technology or trade. This already implies a certain degree of ambiguity with regard to the priorities and future steps of space policy. Meishan Goh (2007: 2) has underlined the link between the parallel maturing of space law, space business, science and technology, and the urgency of an efficient mechanism for dispute settlement. The space market, which, after a phase of expansion, has experienced a phase of consolidation and globalisation, represents great potential for several industrial sectors (such as the building and launching of satellites, satellite operators and service providers in communication, television broadcasting, and localisation; see Bochinger 2009: 33). All these technological and economic interests call for a strong EU space policy and effective legal instruments.

Preliminary remarks on the international framework

To define the scope of the investigation, it is opportune to recall that space law is a branch of international law. In particular, the law of the sea can be considered as the antecedent of air and space law. Indeed, 'the freedom of the high seas would influence both the status of the airspace above the high seas and the freedom of use of outer space' (Haanappel 2003: 11). However, unlike the demarcation between high seas and territorial seas, the distinction between airspace and outer space was not easy. In the nineteenth century, scholars abandoned the united approach in the air domain, inspired by the dichotomy introduced by the international law of the sea (Markoff 1973: 54). The concept of 'sovereign airspace' has been therefore opposed to the 'free outer space', but a consensual definition to identify where the former ends and the latter begins has been lacking (Rosenfield 1979: 137; Oduntan 2003: 64). Furthermore, airspace law may be considered, to some extent, as the antecedent of outer space law, inasmuch as sea navigation shaped air navigation and the latter, in turn, shaped space navigation (Conforti 2013, 282). In light of these considerations, and despite the lack of a specific CJEU case law on outer space, this analysis will focus on certain cases related to airspace to determine whether they may represent a reference for framing outer space (e.g. as a starting point to develop new solutions).

Space law is characterised by the absence of a sector-specific mechanism for dispute settlement, the strong preference to extra-judicial Resolutions (such as arbitration and conciliation), and used to be limited to national judicial disputes (Meishan Goh 2007: 18). Although space law has developed, to some extent, along the lines of the law of the sea, the establishment of the International Tribunal for the Law of the Sea did not lead to the subsequent foundation of a Court specialised in space disputes. Recent trends show that the Member States may accept the renouncement of part of their sovereignty more easily than in the past and admit a compulsory jurisdiction. This is true in several domains where we observe an increasing number of international courts, but not in the field of space law.

If an infringement of international space law obligations comprises a threat to international peace, the United Nations Security Council may intervene. If that is not the case, the general principle of international liability and reparation remains – that is, the obligation of reparation if infringements against international law were committed by a state.

The International Court of Justice (ICJ) has general jurisdiction and universal aspiration, thus disputes concerning international space law can be included in the scope of its competences. However, this jurisdiction has a voluntary nature, which means that the parties have to agree on deferring their disputes to the authority of the ICJ. The ICJ's special jurisdiction, attributed by the parties for a particular dispute, is distinguished by the general jurisdiction, attributed by explicit clauses that may be included in the Treaties with regard to their interpretation and application (Treves 2005: 612).

In addition, specific mechanisms of dispute Resolutions may be provided by international agreements on space, e.g. the Claims Commission established the Convention on International Liability for Damage Caused by Space Objects 1972; again, its decisions will be binding only if its jurisdiction is accepted by the parties (Meishan Goh 2007: 23).

The judges and the advocate generals of the CJEU, which is a regional body, are watching closely the decisions of international courts and dispute settlement bodies in the different domains because the EU and its Member States must respect international law; in this context, the decisions of the international jurisdictions play an important role.

A few cases brought to the ICJ indirectly concern space law, but they are not relevant to this study because they mainly relate to aerial incidents. One case, *Aerial Herbicide Spraying (Ecuador* v. *Colombia)*, presents a different subject.[1] The proceeding was initiated in 2008, but in 2013 it was removed from the Court's list because the parties concerned concluded an agreement that introduced new rules to regulate such activity.

Until now, in the context of the ESA, no dispute under Article XVII of the ESA Convention has been observed. The ESA Council has not yet had the occasion to settle disputes arising between Member States, or between the Member States and the ESA.[2]

Role of the CJEU

The mission of the Court of Justice of the EU, which is composed of the Court of Justice, the General Court (previously Court of First Instance) and the specialised courts (Article 19 Treaty on European Union – TEU) is fairly broad in its scope. The most relevant tasks are assuring the legality of EU legislation, providing the uniform interpretation of EU law and verifying whether the Member States comply with their obligations. Special consideration is, in particular, attributed to the jurisdiction over preliminary rulings under Article 267 TFEU, which 'is regarded as the jewel in the Crown of the existing regime' (Craig 2001: 182), although its importance may substantially vary. In 2013, 64 per cent of the new cases brought before the Court of Justice of the European Union were preliminary rulings (CJEU Annual Report 2013: 84).

In the last few decades, through the different procedures provided by the Treaties, the CJEU became a powerful institution and one of the main pillars of the EU. The Court's case law has strongly contributed to the development of EU law and the progress of EU policies: competition policy and trade policy are clear examples of such a creative and constructive role. The Court has developed fundamental notions, such as the special liability of dominant undertakings

(*ITT Promedia NV* v. *Commission* (1998) ECR II-2937), and has reshaped, to some extent, its own power of control on the European institutions. Scholars have pointed out that 'frames concern power – the power to define and conceptualise' (Mörth 2000: 174) and that in the process of the reformation of frames the interconnections between actors, institutions or policies are reshaped (Surel 2000, 508, Smith 2003, 557). If we admit a broader concept of framing that goes beyond the policy-making in a strict sense, we can assume that the CJEU contributes to framing EU policies.

In the field of trade defence, for instance, the judicial review was expected to be limited because of the highly technical nature of the matter. In contrast, in the EU the General Court has enlarged its competence, especially in anti-dumping cases. First, it addressed the procedural problem of the inadmissibility of the actions for annulment regarding anti-dumping regulations (see, among others, *Euromin SA* v. *Council* (2000) ECR II-02419). Second, it took into consideration substantial issues through an in-depth evaluation of the technical aspects of the trade defence investigations, such as the calculation of the dumping margin. Even the questions falling within the discretion enjoyed by the institutions (as they imply the analysis of complex economic situations) are thus included in the review of the Court, which shall determine whether the procedural rules have been complied with, whether the facts have been accurately stated and whether there has been a manifest error of assessment of those facts or a misuse of power (see, among others, *Detlef Nölle, trading as 'Eugen Nölle'* v. *Hauptzollamt Bremen-Freihafen* (1991) I-05163; *Toyo Bearing and others* v. *Council* (1987) 1809).

In these domains, the progress of European integration, as well as the evolution of related policies, is evident. The study of the CJEU's case law about a specific policy provides important input to evaluate how far such a policy has advanced.

Court of Justice of the European Union case law

Without claiming to be exhaustive, the CJEU's case law under examination is characterised by a certain variety of matters at stake; on the contrary, the type of procedure is, in the majority of cases, the reference to a preliminary ruling (on interpretation or on validity). It thus seems opportune to group the judgements in different categories on the basis of their topics, to identify the solutions provided by the Court and the trend of its orientations.

It is not surprising that the CJEU has mainly intervened where the internal market rules require clarification, in accordance with the so-called 'evolving frames', on the basis of which 'space should not be considered a narrow policy area but one with commercial implications for several EU policy fields' (Stevenson 2012: 91). In spite of the lack of judgements that directly concern outer space, the CJEU's contribution in the domains of air exploitation and dual-use items might indirectly contribute in framing space policy, especially in light of the great potential of the space market. It is true that outer space presents peculiar features distinguishing it from airspace: the interests and values at stake are not identical and similar problems may require different answers. Nevertheless, the Court's findings in the field of airspace will be possibly taken into consideration as a basis from which new solutions will be developed for outer space.

Air exploitation for telecommunication purposes

At the international level, UN Resolution 37/92 established the fundamental principles related to the use of satellites for direct television broadcasting activities (Gorove 1991, 70). As observed by Madders and Thiebaut (1992), the field of telecommunications has been considered particularly important by the EU institutions. The first group of cases under examination concerns the right to transmit television programmes in the air.

In the judgement *Sacchi* (*Giuseppe Sacchi* (1974) ECR 411) the Italian Tribunal referred the case to the CJEU for a preliminary ruling. Under Italian law television was originally a monopoly; Mr Sacchi's television relay undertaking was unauthorised and he refused to pay the licence fee on the television relay receivers. The Court clarified that the television signal should be considered as a provision of service because of its nature, while the sound recordings, the films and the other products for the diffusion of television signals should be regarded as goods. In the opinion of the Court, televised commercial advertising did not come under the provision concerning the adjustment of state monopolies of commercial character (Article 37, European Economic Community Treaty), which referred only to the trade of goods and not the provision of services. As a consequence, the exclusive television right, as such, was not considered in conflict with the EU law. It would have been different if the exclusive right implied a specific benefit for particular trade channels or the benefit of particular commercial operators, compared with others, within the EU.

A few years later, two other preliminary rulings concerned the same domain: 'the diffusion in Belgium of programmes broadcast by television stations outside Belgium' (*Joined Opinion of Mr Advocate General Warner Cases 52/79 and 62/79* (1980) ECR 00833). Because of the geographical configuration of Belgium, many border areas allowed the reception of signals from television stations located in the neighbouring countries. However, although the first case originated in the assumed infringement of the national law prohibiting the diffusion of commercial advertisements (as in *Sacchi*), the second case originated in the assumed breach of copyright (an issue also mentioned in the UN Resolution 37/92, Article 11).

The Belgian broadcasting monopoly was the core of the judgement *Debauve* (*Procureur du Roi* v. *Marc J.V.C. Debauve and others* (1980) ECR 00833). The question regarded the free movement of services and the transmission of commercial advertisements. The nature of service for both the broadcasting of television signals and the transmission of such television signals by cable television was reaffirmed. The Treaties did not preclude national laws from prohibiting the transmission of advertisements by cable televisions, if those laws applied without distinctions with regard to the origin of the advertisements and the service providers. Those rules could therefore be considered neither disproportionate nor discriminating.

In *Coditel* (*Coditel and others* v. *SA Cine Vog Films and others* (1980) ECR 00882) a distribution company obtained by contract the exclusive right to show a specific film in Belgium, both for cinema performance and for television broadcasting. In parallel, the producer assigned to a German television station the right of broadcasting the same film in Germany. A Belgian cable television company picked up the film's signal through its antennas located in Belgium and distributed it to its subscribers. In the opinion of the Court, the free movement of services did not hamper the correct functioning of the intellectual property rules, therefore such activity was unacceptable because of the exclusive rights previously assigned by the film producer to its distributors.

The Court in *Bond van Averteerders and others* (*Bond van Adverteerders and others* v. *the State of the Netherlands* (1988) ECR 02085) referred to the findings of the previous cases, but this dispute was slightly broader in its scope. It concerned certain cable network operators in the Netherlands receiving radio and television programmes supplied from abroad via cable, over the air, or by satellite. As these activities could be qualified as cross-border services, the Treaties' provisions concerning the free movement of services applied. The national law prohibited advertisements in programmes supplied from abroad and provided that programmes with subtitles could only be broadcast with the authorisation of the government (the aim was to exclude commercial advertisements targeting the public of the Netherlands). In other words, as underlined by the Advocate General (*Opinion of Mr AG Mancini, Case 352/85* (1988) ECR 02085), the national law introduced two prohibitions concerning the importation of foreign television programmes. The first was absolute and referred to the advertisements, whereas the second could be overcome through an administrative procedure. The Court considered that these restrictions were in conflict with the free movement of services established by the Treaties and that they were not justified by reasons of public policy.

The 1989 'television directive' (Council Directive 89/552/EEC; Council 1989), which aimed to remove the barriers to the free provision of television broadcasting services, opened the door to another sequence of judgements: two actions for failing to fulfil an obligation and two preliminary rulings.

The first case concerned the UK legislation on broadcasting activities (*Commission of the European Communities* v. *United Kingdom of Great Britain and Northern Ireland* (1996) ECR I-04025). The British law provided certain criteria to determine which satellite broadcasters fell under national jurisdiction and which were considered to be in conflict with the television directive. Moreover, within the British jurisdiction, different regimes applied to domestic satellite services and non-domestic satellite services. A control was exercised over broadcasts, by broadcasters of another Member State, when they were transmitted by a non-domestic satellite service or conveyed to the public as a licensable programme service. In the opinion of the Court, the UK thus failed to fulfil its obligations under the television directive.

In the second case (*Commission of the European Communities* v. *Kingdom of Belgium* (1996) ECR I-04115), the Court examined, among others, a system of prior authorisations for the retransmission by cable of television broadcasts emanating from other Member States for the French-speaking and Dutch-speaking regions. By providing such systems, Belgium failed to fulfil its obligations under the television directive.

The television directive was the object of two preliminary rulings. In *RTI* (*Reti Televisive Italiane SpA and others* v. *Ministero delle Poste e Telecomunicazioni* (1996) ECR I-06471), the Court explained the meaning of the expression 'forms of advertisements such as direct offers to the public', including in this notion the forms of promotion that required more time than spot advertisements on account of their method of presentation. Furthermore, the insertion of the sponsor's name or logo at times other than the beginning and/or the end of the programme was considered allowed. The second preliminary ruling (*Konsumentombudsmannen* v. *De Agostini and others* (1996) ECR I–3875) concerned the compatibility with the EU law of restrictions on television advertising imposed by Swedish law. The Court stated that the television directive and the Treaties did not preclude a Member State from taking measures against television advertising broadcast from another Member State to protect consumers against misleading advertising, provided that those measures did not prevent the retransmission, as such, in its territory of television broadcasts coming from that other Member State. On the other hand, national measures providing that advertisements broadcast in commercial breaks on television must not be designed to attract the attention of children aged less than 12 years of age were considered in conflict with the television directive.

The Court framed the issue of the transmission of television signals via cable, over the air or by satellite. It qualified these activities as provision of services, thus the rules of the internal market related to services apply – in particular, the prohibition of discriminations between domestic and non-domestic. With regard to the derogations, the CJEU recalled that the free movement of services shall not affect the correct functioning of intellectual property rights. These are rulings that may be also applied in future space-dependent services.

Air exploitation for transportation purposes

The second group of judgements concerns a different way of exploiting the air – not for communication, but for transportation purposes. It deals with the aviation sector. Two aspects of the internal market are significant in this context: the protection of customers and the free movement of people. Dempsey and Johansson (2010) outlined the CJEU's approach in the interpretation of Regulation 261/2004 on flight cancellation, delays and denied boarding (European Parliament and Council 2004). They observed that the Court was highly concerned by the interests of air passengers and aimed to give priority to their protection, in light of the Regulation's objectives.

The case *Queen* v. *Department of Transport* (*Queen* v. *Department of Transport* (2006) ECR I-00403) focused, among other issues, on the potential conflict between the Montreal Convention of 1999 (Convention for the Unification of Certain Rules for International Carriage by Air) and EU Regulation 261/2004. A consolidated case law of the CJEU states that the provisions of international agreements prevail over EU secondary legislation. Without prejudice to this principle, the Court affirmed that, in this case, Regulation 261/2004 was not in conflict with the Montreal Convention and that passengers could claim damages on the basis of both legal sources. A distinction was established between the damages that were identical for all the passengers, implying a duty for the air company to take care of them (e.g. providing refreshments), and the individual damages, linked with the personal reasons of the travel. As the EU legislation enhanced the protection provided by the Montreal Convention to air passengers, it could not be considered in conflict with it.

The Court came in for criticism, especially because of the risk of undermining the primary goal of unification, at the international level, of the air carrier liability system (Dempsey and Johansson 2010).

The CJEU's stance was afterwards confirmed (*Wallentin-Hermann* v. *Alitalia* (2008) ECR I-11061). The Court specified the extent of 'extraordinary circumstances' that allowed an exemption from the general duty of compensation and assistance to passengers in the case of denied boarding and of cancellation or a long delay of flights. In particular, mere 'technical problems' were excluded from this notion. Moreover, the CJEU highlighted that the Montreal Convention could not be considered 'decisive' for the interpretation of a notion ('extraordinary circumstances') laid down in EU Regulation No 261/2004, but not in the Convention.

In *Denise McDonagh* (*Denise McDonagh* v. *Ryanair Ltd* ECLI:EU:C:2013:43), the Court provided a further clarification with regard to the closure of part of European airspace as a result of the eruption of the Eyjafjallajökull volcano in Iceland. For the CJEU, in the case of a flight's cancellation caused by 'extraordinary circumstances', such as the volcano's eruption, the obligation to provide care to air passengers should be complied with, although no compensation for the cancellation itself was due.

In this domain a key issue was the distinction between the cancellation and delay of flights. In *Sturgeon* (*Sturgeon* v. *Condor Flugdienst* (2009) ECR I-10923) the Court affirmed that, in the case of delay (even a very long delay) the flight could not be considered cancelled if the other aspects were in accordance with the original plan. Nonetheless, in the case of delays of three or more hours, passengers might rely on the right to compensation, as in the event of cancelled flights, unless the long delay was caused by 'exceptional circumstances'. Through this approach the Court continued to uphold the absence of discordance between the EU legislation and international law. It persisted in shielding the implementation of Regulation 261/2004 for reasons of a higher protection granted to air customers, in particular, in the opinion of Van Dam (2011: 270), 'Sturgeon compensation is compensatory and therefore compatible with the Montreal Convention'. The conclusions of *Sturgeon* were later reaffirmed in the cases *Emeka Nelson and others* v. *Deutsche Lufthansa AG* and *Air France SA* v. *Heinz-Gerke Folkerts* ECLI:EU:C:2012:657: 106.

The Court clarified the notion of 'flight' as 'an air transport operation', and as a consequence a journey out and back was not considered as a single flight (*Emirates Airlines – Direktion für Deutschland* v. *Diether Schenkel* (2008) ECR I-05237).

Moreover, the CJEU specified that, in light of the provisions of Regulation (EC) No. 44/2001 (Council 2001), the Court had jurisdiction to deal with a claim for compensation, on the basis Regulation (EC) No. 261/2004, where the Court had territorial jurisdiction over the place of departure or place of arrival of the aircraft, in the case of the air transport of passengers from one Member State to another Member State, when the contract was concluded with only one airline, i.e. the operating carrier (*Peter Rehder* v. *Air Baltic Corporation* (2009) ECR I-06073).

In the context of air transportation, the emerging frame enshrines the interests of the service's beneficiary over those of the service's provider: higher protection standards, accorded by EU law compared with international law, are endorsed; derogations and exemptions to the carrier's liability are strongly limited. This frame is particularly interesting in the perspective of space tourism, a field where the ESA urges the setting up of a 'regulatory framework adapted to the European scenario' (ESA 2008: 22).

Dual-use technologies

The exportation of dual-use items, which are goods, software and technology that may have both civilian and military purposes, is subject to the control of the EU Commission and to a specific legislation (Council Regulation (EC) No. 428/2009; Council 2009) to avoid any European contribution (voluntary or accidental) to the proliferation of weapons of mass destruction. This policy may affect, to some extent, EU space policy, as numerous space activities and projects may involve dual-use items (for an overview, see Finocchio *et al.* 2008: 3). On the basis of Article 3 of Regulation 428/2009, 'an authorisation shall be required for the export of the dual-use items listed in Annex I', which includes, among others, items concerning the 'aerospace and propulsion' sector (under category 9).

The judgements illustrated are preliminary rulings and actions against Member States for failure to fulfil their obligations. In 1995, in a judgement for preliminary ruling (*Fritz Werner Industrie-Ausrüstungen GmbH and Federal Republic of Germany* (1995) ECR I-03189), the Court confirmed that the licence system represents a quantitative restriction and the Treaties admit as exception to the general principle of free movement of goods national measures aiming to introduce licences for exports of dual-use products. Moreover, the subject of dual-use items must not be excluded in the common commercial policy on the ground that it presents objectives that concern foreign and security policy. It also clarified other important aspects: (1) it is difficult to distinguish between foreign policy and security policy; (2) the concept of public security covers both the internal and external security of a Member State; and (3) the security of a state cannot be considered in isolation.

A second preliminary ruling of the same year allowed a further development of this subject. The Court afterwards reconfirmed the above-mentioned findings (*Peter Leifer and others* (1995) ECR I-03231). The subject of dual-use items is part of the common commercial policy and falls within the exclusive competence of the EU. The Member States enjoy a certain degree of discretion in adopting measures to assure public security, thus they can request to the applicants for export licences to demonstrate that the goods are for civil use; they can refuse the licence if the goods are objectively suitable for military use, and they can issue criminal penalties on the infringements of such rules. The first two measures are considered in accordance with the principle of proportionality. For the third measure, the national authority is competent to decide whether they are proportionate.

Two other cases (*Commission* v. *Italian Republic* (2009) ECR I-11831; *European Commission* v. *Kingdom of Sweden* (2009) I-11777) were actions for failure to fulfil an obligation brought by the European Commission against Sweden and Italy. The Court, in both the judgements, condemned the exemption of imports of dual-use items from customs duties and the refusal to make available to the Commission their own resources, which were not collected because of that exemption. In the conclusions, the two Member States failed to fulfil their obligations. The exceptions provided by the Treaties have to be strictly interpreted and the Member States cannot base their arguments only on the increased price of the imports due to the duties.

In conclusion, the Court examines the interplay of the common commercial policy and the foreign and security policy. On the basis of the frames developed, the issue of dual-use items fell within the common commercial policy, i.e. the exclusive competence of the EU (Article 3 TFEU). Although the discretion of the Member States that is exercised in the adoption of measures of public security is preserved, it cannot be invoked to justify exemptions from their obligations related to the common commercial policy (such as duty obligations). In space policy, the coexistence of commercial and security interests leads to the assumption that this 'nascent frame' (Stephenson 2012: 90) will probably be further developed.

Conclusions

This chapter has highlighted the contribution of the CJEU in framing space policy. The majority of the judgements examined are preliminary rulings and they mainly regard the interpretation of EU law. It has been shown that the CJEU's interpretation activity strongly affects the development of EU policies. Consequently, despite the fact that the CJEU is not a political institution, the study of its frames is useful in assessing how far a given policy has advanced. The analysis of the CJEU case law carried out has allowed some conclusions to be drawn.

First, the lack of a well-established jurisprudence of the CJEU on outer space indicates that EU space policy is still in the early stages. Second, the emerging frames developed by the CJEU related, in particular, to airspace and dual-use items, are indeed relevant. On the one hand, the CJEU's judgements concerning airspace will probably represent a reference for framing outer space, whereas on the other hand the interplay between space policy and other EU policies will probably benefit the CJEU's findings in the domain of dual-use items.

In particular, the exploitation of outer space for communication purposes has to respect the principles of the internal market and, specifically, the rules related to the provision of services. In the perspective of space tourism, the trade-off between the interests of the service's beneficiary and provider could be outlined in light of the high standard of protection assured by the Court to the former. In a sector where trade policy and foreign and security policy are often intertwined, as for the dual-use items, the prerogatives of each policy have been distinguished. If space activities increase, similar questions will be expected to occur.

Although this analysis of CJEU case law cannot be considered decisive in evaluating the development of EU space policy because the number of cases and topics concerned are limited, the sets of frames observed suggest certain possible developments. Moreover, a general frame, representing a sort of guidance, is perceived: the implementation of the internal market for space technologies. Among the possible interests relevant for space policy, those regarding the technological development affecting the trade in goods and the provision of services (in other words, the space business) are likely to be the future priorities. Other interests, linked to exploration and security, for example, will probably take second place.

Notes

1 The information on this case is available from: www.icj-cij.org/docket/index.php?p1=3&p2=3&code=ecol&case=138&k=ee. Last accessed 9 November 2014.
2 A different category of cases concerns the disputes related to the ESA staff or experts and decided by the Appeals Board. Available from: www.esa.int/About_Us/Law_at_ESA/Appeals_Board. Last accessed 9 November 2014.

References

Bochinger, S. (2009) 'Les marchés spatiaux: structure, tendances globales et perspectives'. In: P. Achilleas (ed.), *Droit de l'espace. Télécommunication – Observation – Navigation – Défense – Exploration.* Brussels: Larcier, 33–46.

Conforti, B. (2013) *Diritto internazionale.* Naples: Editoriale Scientifica.

Council (1989) Directive 89/552/EEC of 3 October 1989 on the Coordination of Certain Provisions Laid Down by Law, Regulation or Administrative Action in Member States Concerning the Pursuit of Television Broadcasting Activities. *Official Journal* L 298, 17 October, 23.

Council (2001) Regulation (EC) No. 44/2001 of 22 December 2000 on Jurisdiction and the Recognition and Enforcement of Judgements in Civil and Commercial Matters. *Official Journal* L 12, 16 January 2001, 1.

Council (2009) Regulation (EC) No. 428/2009 of 5 May 2009 Setting up a Community Regime for the Control of Exports, Transfer, Brokering and Transit of Dual-use Items. *Official Journal* L 134, 29 May 2009, 1.

Court of Justice of the European Union (2014) *Annual Report 2013*. Available from: www.curia.europa.eu/jcms/jcms/Jo2_7000/. Last accessed 9 November 2014.

Craig, P. (2001) 'The jurisdiction of the community courts reconsidered'. In: G. de Burca and J. Weiler (eds), *The European Court of Justice*. Oxford: Oxford University Press, 177–214.

Dempsey, P.S. and Johansson S.O. (2010) 'Montreal v. Brussels: the conflict of laws on the issue of delay in international air carriage'. *Air & Space Law*, 35: 207–224.

European Parliament and Council (2004) Regulation (EC) No. 261/2004 of 11 February 2004 Establishing Common Rules on Compensation and Assistance to Passengers in the Event of Denied Boarding and of Cancellation or Long Delay of Flights, and Repealing Regulation (EEC) No. 295/91. *Official Journal* L 46, 17 February 2004, 1.

European Space Agency (ESA) (2008) 'Space tourism – ESA's view on private suborbital space flights'. *ESA Bulletin*, 135. Available from: www.esa.int/About_Us/ESA_Publications/ESA_Publications_Bulletin. Last accessed 9 November 2014.

Finocchio, P., Prasad, R. and Ruggieri, M. (2008) *Aerospace Technologies and Applications for Dual Use*. Aalborg: River Publishers.

Gorove, S. (1991) *Developments in Space Law – Issues and Policies*. Boston, MA: Martinus Nijhoff.

Haanappel, P.P.C. (2003) *The Law and Policy of Air Space and Outer Space*. The Hague: Kluwer Law International.

Madders, K.J. and Thiebaut, W.H. (1992) 'Two Europes in one space: the evolution of relations between the European Space Agency and the European Community in space affairs'. *Journal of Space Law*, 20: 117–133.

Markoff, M.G. (1973) *Traité de Droit International Public de l'espace*. Fribourg: Éditions Universitaires Fribourg Suisse.

Meishan Goh, G. (2007) *Dispute Settlement in International Space Law*. Boston, MA: Martinus Nijhoff.

Mörth, U. (2012) 'Competing frames in the European Commission – the case of the defence industry and equipment issue'. *Journal of European Public Policy*, 7(2): 173–189.

Oduntan, G. (2003) 'The never ending dispute: legal theories on the spatial demarcation boundary plane between airspace and outer space'. *Hertfordshire Law Journal*, 1(2): 64–84.

Rosenfield, S.B. (1979) 'Where airspace ends and outer space begins'. *Journal of Space Law*, 7: 137–148.

Smith, M. (2003) 'The framing of European foreign and security policy: towards a postmodern policy framework?' *Journal of European Public Policy*, 10(4): 556–575.

Stephenson, P. (2012) 'Talking space: the European Commission's changing frames in defining Galileo'. *Space Policy*, 28: 86–93.

Surel, Y. (2000) 'The role of cognitive and normative frames in policy-making'. *Journal of European Public Policy*, 7(4): 495–512.

Treves, T. (2005) *Diritto internazionale – problemi fondamentali*. Milan: Giuffré.

United Nations (1982) *Resolution 37/92, 10 December 1982, Principles Governing the Use by States of Artificial Earth Satellites for International Direct Television Broadcasting*. Available from: http://research.un.org/en/docs/ga/quick/regular/37. Last accessed 9 November 2014.

Van Dam, C. (2011) 'Air passenger rights after Sturgeon'. *Air & Space Law*, 36: 259–274.

Court of Justice of the European Union: table of cases

Case 155/73, *Giuseppe Sacchi* (1974) ECR 411.

Opinion of Mr Advocate General Reischl, Case 155/73 (1974) ECR 411.

Case 52/79, *Procureur du Roi* v. *Marc J.V.C. Debauve and others* (1980) ECR 00833.

Case 62/79, *SA Compagnie Generale pour la Diffusion de la Television, Coditel, and others,* v. *SA Cine Vog Films, and others* (1980) ECR 00882.

Joined opinion of Mr Advocate General Warner, Cases 52/79 and 62/79 (1980) ECR 00833.

Case 240/84 NTN, *Toyo Bearing and others* v. *Council* (1987) ECR 1809.

Case 352/85, *Bond van Adverteerders and others* v. *the State of the Netherlands* (1988) ECR 02085.

Opinion of Mr Advocate General Mancini, Case 352/85 (1988) ECR 02085.

Case C-16/90, *Detlef Nölle, trading as 'Eugen Nölle'* v. *Hauptzollamt Bremen-Freihafen* (1991) ECR I-05163.

Case C-70/94, *Fritz Werner Industrie-Ausrüstungen GmbH and Federal Republic of Germany* (1995) ECR I-03189.

Case C-83/94, *Peter Leifer, Reinhold Otto Krauskopf, Otto Hölzer* (1995) ECR I-03231.

Case C-222/94, *Commission of the European Communities* v. *United Kingdom of Great Britain and Northern Ireland* (1996) ECR I-04025.

Opinion of Mr Advocate General Lenz, Case C-222/94 (1996) ECR I-04025.

Joined cases C-320/94, C-328/94, C-329/94, C-337/94, C-338/94 and C-339/94, *Reti Televisive Italiane SpA (RTI)* (C-320/94), *Radio Torre* (C-328/94), *Rete A Srl* (C-329/94), *Vallau Italiana Promomarket Srl* (C-337/94), *Radio Italia Solo Musica Srl and others* (C-338/94) and *GETE Srl* (C-339/94) v. *Ministero delle Poste e Telecomunicazioni* (1996) ECR I-06471.

Opinion of Advocate General Jacobs, Joined cases C-320/94, C-328/94, C-329/94, C-337/94, C-338/94 (1996) ECR I-06471.

Joined Cases C-34/95, C-35/95 and C-36/95, *Konsumentombudsmannen* v. *De Agostini, and Konsumentombudsmannen (KO)* (C-34/95) and *TV-Shop i Sverige AB* (C-35/95 and C-36/95) (1996) ECR I – 3875.

Case C-11/95, *Commission of the European Communities* v. *Kingdom of Belgium* (1996) ECR I-04115.

Opinion of Mr Advocate General Lenz, Case C-11/95 (1996) ECR I-04115.

Case T-111/96, *ITT Promedia NV* v. *Commission* (1998) ECR II-2937.

Case T-597/97, *Euromin SA* v. *Council* (2000) ECR II-02419.

Case C-344/04, Queen v. Department of Transport (2006) ECR I-00403.

Opinion of Mr Advocate General Geelhoed, Case C-344/04 (2006) ECR I-00403.

Case C-549/07, *Wallentin-Hermann* v. *Alitalia* (2008) ECR I-11061.

Case C-173/07, *Emirates Airlines – Direktion für Deutschland* v. *Diether Schenkel* (2008) ECR I-05237.

Case C-204/08, *Peter Rehder* v. *Air Baltic Corporation* (2009) ECR I-06073.

Joined Cases C-402/07 and C-432/07, *Sturgeon* v. *Condor Flugdienst* (2009) ECR I-10923.

Opinion of Advocate General Sharpston, Joined Cases C-402/07 and C-432/07 (2009) ECR I-10923.

Case C-387/05, *European Commission* v. *Italian Republic* (2009) ECR I-11831.

Case C-294/05, *European Commission* v. *Kingdom of Sweden* (2009) I-11777.

Joined cases C-581/10 and C-629/10, *Emeka Nelson and others* v. *Deutsche Lufthansa AG,* and *TUI Travel plc and others* v. *Civil Aviation Authority* ECLI:EU:C:2012:657.

Opinion of Mr Advocate General Bot, Joined cases C-581/10 and C-629/10, ECLI: EU:C:2012:295.

Case C 12/11, *Denise McDonagh* v. *Ryanair Ltd*, ECLI:EU:C:2013:43.

Case C-11/11, *Air France SA* v. *Heinz-Gerke Folkerts, Luz-Tereza Folkerts*, ECLI: EU:C:2013:106.

Part III
Politics
Internal and external perspectives

Part III

8 Satellites and farming

Reframing agriculture and agricultural machinery within EU space policy

Ulrich Adam

Massive investments in agricultural research and technology will be required in the coming years to accelerate global food production and to feed the world affordably and sustainably. According to the United Nation's Food and Agriculture Organization (FAO), 'an average net investment of 83 billion dollars a year will be necessary to raise agricultural production by 60 percent and feed the global population of more than 9 billion expected by 2050'.[1] The search for innovative technologies that will enable farmers to live up to this challenge has intensified. The question of which technology farmers should embrace remains hotly contested, given the wide-ranging, worldwide effects of those choices. As the Washington-based think-tank, the International Food Policy Research Institute (IFPRI), noted in its 2014 report, '[f]uture choices and adoption of agricultural technologies will fundamentally influence not only agricultural production and consumption but also trade and environmental quality' (IFPRI 2014: 3).

To find technology solutions that can raise productivity and output levels in farming while also promoting sustainability and greater environmental protection has proved to be particularly challenging. Farming by satellite – commonly termed precision agriculture – holds the promise of boosting both sustainability and productivity levels in farming. As such, it has emerged as a strategically important technology solution that can help to meet the challenge of the growing world population in the run-up to 2050. Space policy and the rapid growth of satellite technology applications in agricultural machinery have presented the European agricultural machinery industry with a unique opportunity to transform adverse and outdated views about agriculture and agricultural machinery in EU policy-making.

To understand this fundamental shift of frames it is important to start by considering the highly adverse views that farming and agriculture-related operators such as the agricultural machinery industry faced at the EU level until around 2008. Before then, agriculture was widely seen as a rather outdated, backward sector, particularly when compared with the seemingly more dynamic, knowledge-intensive industries such as the information technology or renewable energies sectors. This perception was underpinned by the fact that the agricultural sector had been suffering from a long-standing decline in terms of overall economic value and direct employment. In France, for instance, employment in agriculture had fallen from 32 per cent in 1952 to a mere 3 per cent in 2012 (Piketty 2014: 91).

At the EU level, agriculture and farming have always been perceived in the context of a policy frame that has been controversially discussed given its interventionist nature and substantial budget implications: the Common Agricultural Policy (CAP) (Lynggaard 2007). To make matters worse, wherever and whenever forward-looking technology debates on the future of farming have emerged at the EU level in recent years, they have tended to become quickly locked up in a bipolar frame. At one end of the spectrum, advocates of low-input and organic agriculture have called for farming approaches that prioritise environmental protection over productivity growth and have rejected technological solutions that they consider would go against this goal (IFOAM EU 2013). At the other end of the spectrum, advocates of intensive agriculture have assumed that high levels of agricultural inputs, together with the help of new – and often controversial – technology approaches such as genetic modification, hold the key to the future of farming (EASAC 2013).

Against this background, EU space policy and the rapidly growing importance of satellite technology applications in farming have presented the agricultural machinery industry with the opportunity to establish space as a frame within European agricultural policy and, with the help of this move, to turn around outdated views and beliefs about farming and the industry in Europe. This chapter analyses the different aspects of this remarkable reframing exercise. After a brief overview of the role of space technology applications in modern day farming and a look at the market opportunities they present, this chapter outlines how the positive connotations linked to space policy have been used by the European agricultural machinery industry:

- to underline the broader societal relevance and strategic importance of farming by satellite technology to feed a world of more than nine billion people by 2050;
- to portray modern day farming as a cutting-edge, knowledge-intense activity that is undergoing a technological revolution;
- to demonstrate how precision farming equipment can cut through the deadlock between traditionalist and progressive voices in farming and present a widely accepted technology solution that helps to reconcile the quests for greater sustainability and higher productivity in agriculture.

In doing so, this chapter also considers the roles that related developments have played in changing perceptions and how far they have structurally helped to support – or slow down – the change. The chapter closes with a discussion of the future opportunities and risks that could result from this new perception.

Frames

Technology frame: farming by satellite

In agriculture, high-accuracy satellite-based positioning systems are the entry point for what is commonly described as precision agriculture or precision farming. By providing farmers with navigation and positioning capability any-where on earth, at any time and under all conditions, satellite-based systems help to achieve unprecedented levels of accuracy and precision when driving and working in the field. To achieve this level of accuracy, a standard Global Naviga-tion Satellite System (GNSS) receiver is complemented by free satellite-based augmentation services, such as the European Geostationary Navigation Overlay Service (EGNOS), which will soon be replaced by Galileo. When these are com-bined with additional ground-based infrastructure systems, such as real-time kin-ematics or the differential global positioning system, they can locate and navigate agricultural machines on a field with an accuracy of 2 cm. As a result of such automatic systems, farmers can drive on perfectly parallel lines in the field, avoiding overlaps and gaps in cultivation activities. Autopilot applications also reduce driver fatigue and improve the quality of farming processes (van der Waal 2014). Major satellite-based applications in agriculture include:

- automatic steering systems for tractors (or combine harvesters) that take over the steering so that the farmer can take his or her hands off the wheel and focus fully on monitoring and managing the overall working process;
- variable rate technology, in which precise location information is used to tailor farm inputs (such as water, seeds, fertilisers or pesticides) to the exact needs on the ground (e.g. according to the level of soil dryness, quality, plant growth or plant health);
- fleet management and documentation services that enable farmers to monitor the exact location of different machines involved, for instance, in the harvesting process to coordinate their tasks remotely and record all their activities.

Market opportunity frame

The new possibilities in farming offered by satellite-based technologies have been recognised by the EU and the European industry as an immense market opportunity. In line with this, market opportunity and the promotion of European industrial competitiveness have always been two fundamental benefits used by the Commission to justify the expenses for Galileo (Stephenson 2012: 86). This could be observed in the inauguration speech of European Commission President Jean-Claude Juncker at the European Parliament in October 2014, during which he emphasised that '[s]pace policy can make an important contribution to the further development of a strong industrial basis in Europe', reiterating space pol-icy's prominent standing among the planned EU initiatives in the area of 'indus-trial policy' under the EU's long-term economic strategy, the EU 2020 Strategy.[2] Another European Commission representative, Matthias Petschke, Director for EU Satellite Navigation Programmes, summarised the industrial importance of space policy for the EU in 2013 as follows: 'We rely on satellite technologies every day', hence satellite technology 'is an essential pillar of the EU's strategy to enhance competitiveness for jobs and growth'.[3]

The overall market opportunity for space technologies in agriculture – in terms of volume and value – is relatively small compared with other sectors. According to the latest market forecasts, agricultural applications will represent only 1.4 per cent of the overall revenue expected from satellite navigation technology sales between 2012 and 2022 (European Global Satellite Navigation Systems Agency 2012, 2013). Yet the market dynamics in farming are extremely favourable. The growth of space technology applications in agriculture is predicted to be exceptionally strong, truly global and poised to set off a virtuous cycle because falling prices for satellite technology equipment will trigger a greater demand and investment among farmers in return.

These positive market developments are confirmed in the regular market reports of the European GNSS Agency, which provides detailed overviews of the expected economic benefits in terms of the retail value of GNSS receivers, maps and the navigation software used in agriculture. According to the latest reports, the market growth of GNSS device sales in agriculture is predicted to be above average (16 per cent per annum until 2020). In the Netherlands, for instance, the use of GNSS tools skyrocketed from a mere 10 per cent of arable farms in 2007 to 65 per cent just six years later in 2013 (van der Waal 2014). Globally, annual sales are set to quintuple from about 100,000 units in 2010 to more than 500,000 units in 2020, at which point a penetration level of approximately 40 per cent will be met. Dynamic growth can be observed across all regions – Western Europe, Central and Eastern Europe, and the emerging economies. Moreover, the pricing dynamics are favourable, as both the prices of high-end and low-end GNSS equipment are predicted to fall, making it possible for increasing numbers of farmers to invest in GNSS equipment.

Global food security frame: how to feed a world of ten billion people by 2050

To use a frame such as space prominently in political agenda-setting, it is essential to demonstrate broader societal relevance and strategic importance and thus to move beyond simply hard economic evidence. Apart from the clear market opportunity for satellite technology applications, EU policy-makers have been quick to point out that space policy 'can provide the tools to address many of the global challenges that face society in the twenty-first century'.[4] Among the global challenges listed in this context are: climate change, air pollution, the management of natural resources and desertification. The 2008 and 2011 spikes in food commodity prices added another important issue to this list: global food security.[5] Even in the best of cases, achieving the level of agricultural productivity needed to meet the rapidly rising world demand for food, fibre and fuel by 2050 will be a tremendous challenge. This challenge is set to become even more daunting because it will be aggravated by a number of new constraints, such as the gradual slow-down in the growth of agricultural yield, the limited availability of new arable land, the effects of climate change, the price and availability of energy, and the negative impacts of urbanisation on the rural labour supply. The combined effect of these constraints has been the reason for the FAO to label the global food security challenge as 'a gathering storm'. As a result of these developments, agriculture and farming are suddenly back on the highest political agenda – and, with them, the search for innovative technologies that can empower farmers and the farming community to live up to this challenge. As the FAO acknowledged in this context, satellite-based farming technologies, such as precision irrigation or precision fertilising techniques, could make an important contribution to growing more food with a lower input of resources (FAO 2011).

Discussion

Reconciling the irreconcilable: farming by satellite and the sustainable intensification of agriculture

Following the emergence of the global food security debate, debates on the future of farming in Europe have tended to focus on different and potentially conflicting sub-frames. The first of these is the productivity frame – that is, the search for ways to increase crop yields and drive up outputs – a task seen by many as the primary task in agriculture today. The second is the sustainability frame – that is, the growing societal quest for greater sustainability and environmental protection in farming. The final frame is that on the economic feasibility of farming – that is, the quest of farmers to make sure that they can continue to make an appropriate living from their activities.

Debates on the introduction of new technologies in farming have tended to become quickly locked up between these sub-frames. In fact, in recent years, a veritable turf war has erupted over them. Finding technology solutions that can truly reconcile greater productivity and sustainability in farming has proved particularly difficult. Some debates – such as the debate between the proponents of intensive agriculture on the one hand and the advocates of organic farming on the other – seem to suggest that both frames are mutually exclusive and that we will ultimately be forced to make a choice for one side or the other. Another prominent example of a debate that has largely been fought over the alleged irreconcilability between the productivity and sustainability frames in Europe is the debate over the use of genetically modified plants in agriculture. Although proponents have pointed out the immense potential of genetically modified plants in making agricultural crops more productive (e.g. as a result of greater drought or pest resistance), opponents view the risks for the environment as uncontrollable and thus diametrically opposed to any notion of their concept of sustainability in farming. Similar debates over the use of other key inputs in modern agriculture, such as pesticides and fertilisers, have developed largely along the same lines.

As a result, many policy-makers have avoided positioning themselves too clearly on either side of the productivity or sustainability frames and have instead tried to present their own agricultural policy approaches and strategies as a blend of both. In line with this, the FAO simply decided to mix both frames in their first strategic objective that focuses on sustainable crop production intensification

(SCPI) in agriculture. The FAO defines SCPI as producing more from the same area of land while reducing negative environmental impacts. However, such a move does not necessarily provide a clear answer on how exactly the SCPI challenge will be met – that is, which of the many controversially discussed technology solutions should be pursued to achieve it. This is where precision agriculture and farming by satellite have emerged in a new, prominent and particularly favourable light. Against the background of the global food security challenge and the apparently contradictory frames on the future of farming, farming by satellite can be seen and portrayed as a solution that could truly help farmers to achieve the SCPI objectives and successfully reconcile the productivity and sustainability frames with the help of one common technology approach.

Space technology in agriculture could also bring a number of further considerable benefits. By allowing farmers to save on input costs and operate in a more efficient manner, farmers using satellite technology could benefit economically. They would also enjoy much greater working comfort as a result of the higher levels of machinery automation that come with it. Satellite technology could enable an unprecedented level of documentation in agriculture as a result of the continuous collection of farm data. EGNOS, for instance, has already been used within the CAP for on the spot checks to verify subsidy claims. As such, farming by satellite promises to deliver hitherto unknown levels of transparency and thus consumer protection in agricultural operations. The broad consensus about the benefits of farming by satellite is best summarised in the words of the European Commission's EGNOS unit, which notes that '[s]atellites contribute to improve the quality and strengthening the safety of our food, while also protecting the environment' (European Commission 2011: 5).

As a result, the space frame in agricultural policy has quickly achieved a much higher societal acceptance than, for instance, the much more contested biotechnology or bioeconomy frames have managed to achieve (Daviter 2012). Even staunch adversaries of biotechnology and industrial farming acknowledge that farming by satellite can bring real benefits. For instance, the leading bioeconomy critic Franz-Theo Gottwald notes in his latest book that, thanks to precision farming and farming by satellite, farmers are able to 'comply in a more reliable manner with environmental directives without loss of turnover. Thanks to IT-supported documentation systems that come with it, consumer protection and the demand for transparency are also taken into account' (Gottwald and Krätzer 2014: 36)

Making farming smart, productive and sustainable: the agricultural machinery industry as an advocate of farming by satellite and EU space policy

Ever since the emergence of space as a major technological solution to some of the most fundamental challenges in agriculture, the European agricultural machinery industry has been quick to promote and establish it in agricultural policy. Previously, the industry had primarily taken recourse to market- and technology-related arguments to appeal to EU policy-makers and present their interests. With an annual production value of €28 billion and 4500 companies active in the sector, the agricultural machinery industry has always been quick to emphasise that it is the world's largest producer of farm equipment. In addition, the industry has successfully pointed out that agricultural machinery is one of the most innovative industry sectors in Europe. This was confirmed in the European Commission's competitiveness report published in October 2013, in which agricultural machinery featured as the second highest ranking European industry sector in terms of innovation and competitive advantage (European Commission 2013: 33).

The ability to use the space frame made the prominent positioning of the agricultural machinery industry in political agenda-setting considerably easier and clearer. First, it allowed the industry to enter into the broader debate on global food security by emphasising how precision farming technology could promote the sustainability, productivity and economic viability of farming. Second, the inherent flexibility and multi-dimensional nature of the space frame enabled the industry to use it successfully with different emphasis and in different forms with different actors in EU policy debates on agriculture.

In line with this, one immediate objective of the agricultural machinery industry has been to gather and communicate facts and evidence about the practical benefits of farming by satellite and smart machinery to ensure that these are better understood by EU policy-makers. The industry has tried to achieve this through a variety of activities, starting with the organisation of an exhibition at the European Parliament on 'Smart equipment for sustainable agriculture – precision farming: producing more with less' which took place in October 2013.[6] The exhibition spelled out the specific contributions that different types of machines can deliver in terms of:

- agricultural productivity gains (e.g. increases in yield, improvements in crop quality);
- greater sustainability (e.g. reduced use of inputs, reduced environmental footprint of agricultural operations through tailored application processes);
- improved economic viability of farming (higher profitability as a result of lower input costs).

To look at a specific example, the exhibition showed that precision fertilising equipment can deliver typical increases in yield of 6–8 per cent while reducing fertiliser application rates by up to 15 per cent. As a result, farmers can achieve effective savings of up to €100 per hectare of land.

The data and information gathered for the exhibition also supported the more long-term advocacy goal of the industry to work towards an integrated and coherent EU policy approach to precision farming – that is, to ensure greater alignment between the different Directorates-General within the European Commission that deal with this issue. The Commission is responsible for a broad variety of EU programmes and initiatives that – directly or indirectly – touch upon smart equipment and farming by satellite. These range from programmes for satellite navigation technology (in the form of EGNOS, Galileo and Copernicus) to research funding (Horizon 2020), initiatives to promote the practical use of precision farming equipment (the European Innovation Partnership on Agricultural Productivity and Sustainability) to subsidy schemes for farmers under the CAP. Looking at Horizon 2020, the industry has actively started to participate in research consortia that have submitted proposals under the newly created funding line for agricultural research. Representatives from the agricultural machinery industry have been appointed to the new Focus Group on Precision Farming under the European Innovation Partnership on Agricultural Productivity and Sustainability, which advises the European Commission on future research priorities.

In addition, the industry provided a systematic input into the analytical report that was published by the European Commission's Joint Research Centre for the European Parliament in July 2014. The report confirmed the substantial role of farming by satellite and precision equipment 'in meeting the increasing demand for food while ensuring sustainable use of natural resources and the environment' (Joint Research Centre of the European Commission 2014). Most importantly, it called for further EU action such as 'awareness-raising and information campaigns among farmers, the provision of appropriate guidelines, and an EU "precision farming calculator" tool which would bring decision-support value to farmers and advisers'. All this should be accompanied by 'research and development studies, for instance to assess the impact of precision agriculture on the environmental footprint beyond the farm level' (Joint Research Centre of the European Commission 2014).

Another prominent example of the industry's use of the space frame in the context of advocacy at the EU level is the Farming by Satellite Prize. Established in 2012 by the European GNSS Agency under the auspices of the EU, the prize is financially supported by an agricultural machinery company as well as other stakeholders in the agricultural production chain. All students and young people aged less than 32 years are invited to participate and to propose innovative ideas for the use of satellite services in agriculture, food production or land management. The aim of the prize is firmly described in terms of market opportunity and technology sub-frames – that is, to help European entrepreneurs and businesses commercially exploit EGNOS and Galileo and to ensure that European industry maintains a competitive edge in the global satellite navigation technology marketplace. However, the primary aim of the prize – to promote the benefits generated by the GNSS in agriculture among farmers – is firmly described in terms of the global food security and sustainable intensification of agriculture frames: 'to promote the use of the [GNSS] in agriculture and its benefits to farmers, consumers, food security and geo-traceability, remote sensing technologies, sustainable land management, and the environment'.[7] The Farming by Satellite Prize helped to change the image of the agricultural machinery industry towards graduates and students by showing it as a dynamic, high-technology sector and thus as an attractive future career choice.

As can be seen by these examples, the space frame has enabled the agricultural machinery industry to provide input into a much broader range of EU legislative initiatives than seemed previously possible. Most importantly, it has enabled the industry to receive greater political attention and forge a closer dialogue with decision-makers by emphasising its strategic importance and solution potential in the context of the global food security challenge.

Future outlook: opportunities and risks

There have been many structural benefits to the agricultural machinery industry of using space as a frame in political agenda-setting at the European level. This is likely to continue in the foreseeable future. High-level events such as EXPO 2015 in Milan and political debates on the upcoming mid-term review of the CAP will provide further opportunities to use the space frame in agricultural policy debates. With its umbrella theme of 'Feeding the Planet, Energy for Life', EXPO 2015 will put a strong focus on technology solutions that will help to make farming more productive and sustainable in the coming years. Smart machinery and farming by satellite will feature prominently among them Aspects related to precision farming will be displayed in the EU's EXPO Pavilion and at a one-day conference organised by the European Commission in Milan. In addition, the mid-term review of the CAP will once again see the political debate on technology solutions for sustainable agricultural productivity growth intensify at the EU level. As the 2014 report of the Joint Research Centre of the European Commission highlighted, '[p]romoting precision agriculture through the CAP seems to be economically, environmentally and even socially justifiable' (Joint Research Centre of the European Commission 2014: 43). It remains to be seen exactly how the CAP could further be adapted to better support the advancement and uptake of precision agriculture in Europe.

So what about untapped opportunities or the potential risks of using space as a frame in agricultural policy debates? In the long run, the current focus on space and satellite technologies in agricultural policy is likely to be complemented by further new technology approaches. One such concept could be that of robotics. Already well-established in the form of feeding and milking robots in livestock farming, the use of robots is still in its infancy in arable crop farming. However, they are set to hit the fields in the years ahead and will allow for an even better and more targeted management of farm operations. As such, robots will fit the sustainability and productivity frames in the same way that farming by satellite does today. For instance, a weeding robot could be used to help keep agricultural productivity high, yet will also promote sustainability in the field as a result of the reduced use of pesticides.

So what about the potential risks and limitations of using space as a frame in agriculture? Overall, it is remarkable that there seem to be so few. The greatest limitation regarding the use of satellite technology in farming is of a technical nature and relates to the challenge of improving the interoperability and interconnectivity between the different navigation systems and machinery components available on the market. Another major limitation is financial, as it is still challenging for a number of farmers in Europe to purchase and use high-end satellite technology on their farms. Again, the debate is likely to focus on the question of how the CAP could structurally support farmers in purchasing and using high-end satellite technology in the future.

A more serious future challenge for precision agriculture could be data protection. As modern farm tractors have become data-rich sensing machines that constantly collect and transmit machine-specific and agronomic data from the field (e.g. the status of plant growth, plant health and projected harvest), it will be important to ensure that questions regarding secure data transfer and data ownership are answered convincingly and adequately regulated. This will be particularly important as the use of cloud-based farm management systems is increasing. At the same time, it is evident that the question of data protection can and will not be answered within the remit of farming and farm machinery alone. As such, the question of whether data protection will become an issue in farming will largely depend on how the broader public debate on this issue continues to evolve at the European level.

Conclusion

Space policy and space technology applications have played a major role in reframing the image of farming and the agricultural machinery industry in European political agenda-setting in recent years. In light of the many opposing frames that agriculture and the agricultural machinery industry were facing until around 2008, it is fair to conclude that the transformative story that agriculture and the agricultural machinery industry has undergone in recent years would not have been as successful without the use of the space frame. As this chapter has shown, there are a number of reasons that can explain why space could be introduced so quickly, firmly and successfully to existing discussions on agriculture and farm machinery. First, space and space technology applications in farming could be presented as a tremendous market opportunity for the European industry. As a result of the positive connotations linked to space and space technology, farming with advanced machinery could be portrayed in an entirely new light as a cutting-edge, knowledge-intense activity undergoing a fundamental technological change.

However, by themselves, these factors would probably have been insufficient to trigger the broader change in perception described in this chapter. The reason why space and precision farming could be used in political advocacy in such a compelling way is because they allowed the industry to become involved and profile itself in a clearer, more prominent and wider way.

In terms of clarity and prominence, space allowed the industry to become directly involved in the much broader debate on global food security and the future of farming. By presenting farming by satellite as a solution to a worldwide challenge – that is, the question about which technology solutions could and should be used to drive up productivity levels in agriculture in ways that will make it possible to feed a world of ten billion people sustainably by 2050 – the industry was able to explain that farming and farm machinery had become strategically important issues of wider societal relevance. In addition, it enabled the agricultural machinery industry to showcase precision farming equipment as a solution that could cut through the deadlock between traditionalist and progressive voices in farming and to present a widely accepted technology solution for the sustainable intensification of agriculture.

In terms of a wider involvement in EU policy debates, the space frame has enabled the industry to become involved in a much broader range of EU programmes and initiatives. Apart from space policy proper, this concerns, in particular, EU policies in the areas of research and agriculture. As the current risks of using the space frame in agriculture appear to be low, further untapped opportunities seem to be plentiful for the foreseeable future.

Acknowledgements

The author thanks the editors for their feedback and help in writing this chapter. This article was written by the author in a personal capacity.

Notes

1 Quoted from: Elena L. Pasquini, '4 takeaways from the 2014 Committee on Food Security', *Devex*, 20 October 2014. Available from: www.devex.com/news/4-takeaways-from-the-2014-committee-on-food-security-84587.
2 'Setting Europe in Motion', President-Elect Juncker's main messages from his speech before the European Parliament, 22 October 2014. Available from: europa.eu/rapid/press-release_SPEECH-14-705_en.htm.
3 'European Space Solutions: Benefits and Opportunities', Report from the European Space Solutions Conference held in Munich 5–7 November 2013. Available from: www.gsa.europa.eu/news/european-space-solutions-benefits-and-opportunities.
4 Quoted from: www.ec.europa.eu/enterprise/policies/space/index_en.htm.
5 For an introduction to the renewed debate on global food security, see Barrett (2013) and De Castro *et al.* (2013).
6 The online version of CEMA's exhibition has been published at www.cema-agri.org/page/cema-exhibition-%E2%80%9Csmart-equipment-sustainable-agriculture%E2%80%9D.
7 Quoted from www.farmingbysatellite.eu/press/overview-of-the-prize/.

References

Barrett, C.B. (ed.) (2013) *Food Security and Sociopolitical Stability*. Oxford: Oxford University Press.
De Castro, P., Adinolfi, F., Capitanio, F., Di Falco, S. and Di Mambro, A. (eds) (2013) *The Politics of Land and Food Scarcity*. London: Routledge.
Daviter, F. (2012) *Framing Biotechnology Policy in the European Union*. ARENA Working Paper 05/2012. Oslo: ARENA Centre for European Studies.
European Academies Science Advisory Council (EASAC) (2013) *Planting the Future: Opportunities and Challenges for Using Crop Genetic Improvement Technologies for Sustainable Agriculture*. EASAC Policy Report 21, June 2013. Brussels: EASAC.

European Commission (2011) *Space for Citizens: Bringing Space down to earth.* Luxembourg: Publication Office of the European Union.

European Commission (2013) *Towards Knowledge Driven Reindustrialisation.* European Competitiveness Report 2013. Luxembourg: Publication Office of the European Union.

European Global Satellite Navigation Systems Agency (2012) *GNSS Market Report – Issue 2* [online]. Available from: www.gsa.europa.eu/system/files/reports/Market%20 Report%20MEP7%202012%20WEB_0.PDF. Last accessed 14 May 2015.

European Global Satellite Navigation Systems Agency (2013) *GNSS Market Report – Issue 3* [online]. Available from: www.gsa.europa.eu/system/files/reports/GSA%20 -Market%20Report%202013%20new_1.pdf. Last accessed 14 May 2015.

Food and Agriculture Organisation (FAO) of the United Nations (2011) *Putting Nature Back into Agriculture* [online]. Available from: www.google.co.uk/url?sa=t&rct=j&q= &esrc=s&source=web&cd=1&ved=0CCIQFjAA&url=http%3A%2F%2Fwww.fao.org %2Fnews%2Fstory%2Fen%2Fitem%2F80096%2Ficode%2F&ei=XU5PVZu2N4jdUd GngfAI&usg=AFQjCNHuk7yu_akddHsQ3DzzkBlKw-EjSA&bvm=bv.92885102,d. d24. Last accessed 10 May 2015.

Gottwald, F.-T. and Krätzer, A. (2014) *Irrewg Bioökonomie. Kritik and einem totalitären Ansatz.* Berlin: Suhrkamp.

IFOAM EU (2013) *Making Europe More Organic – 10 Years of Advocacy for Sustainable Food and Farming.* Brussels: IFOAM EU Group.

International Food Policy Research Institute (IFPRI) (2014) *Food Security in a World of Natural Resource Scarcity: The Role of Agricultural Technologies.* Washington, DC: IFPRI.

Joint Research Centre of the European Commission (2014) *Precision Agriculture: An Opportunity for EU Farmers – Potential Support with the CAP.* Luxembourg: Publications Office of the European Union.

Lynggaard, K. (2007) 'The institutional construction of a policy field: a discursive institutional perspective on change within the common agricultural policy'. *Journal of European Public Policy*, 14(2): 293–312.

Piketty, T. (2014) *Capital in the Twenty-First Century* (tr. A. Goldhammer). Cambridge, MA: Harvard University Press.

Stephenson, P. (2012) 'Talking space: the European Commission's changing frames in defining Galileo'. *Space Policy*, 28(2): 86–93.

van der Waal, T. (2014) 'Seeds of growth'. *Geo International*, May: 20–22.

9 Space policy in the context of trans-European networks and the completion of the Single Market

Paul Stephenson

Introduction

This chapter examines the development of the trans-European networks (TENs) and the case for investing in common infrastructures as a tool for growth. It posits that the emergence of EU space policy programmes such as Galileo must be understood in the context of developments in large European infrastructure projects in the fields of transport, energy and telecoms, first proposed in 1990 (Commission 1990a) and agreed at the Essen European Council in December 1994 (European Parliament and Council 1996). The first revision of the TEN-T list of priority projects in 2004 saw it expanded from 14 to 30 priority projects (European Parliament and European Council 2004). By incorporating the Galileo project within TENs, satellite navigation became an integral part of EU transport policy. The applications derived from Galileo were to be relevant to traffic management in air, rail and maritime transport. More specifically, satellite navigation was intended to enhance the efficient operation of rail freight services, internal waterways traffic and short-sea shipping. Space policy was framed as a driver of market integration, contributing to the further liberalisation of services and effective monitoring of passenger and freight movements, looking forward to 2030 and beyond.

Arguably, without the advances first made in policy-making for terrestrial infrastructures, there would not have been the political climate or consensus to invest in common space infrastructures. It is important to analyse the context of TENs and how they were framed to fully understand the development of Galileo over time. Time and sequence are essential considerations in any analysis of European policy-making. European space policy must be understood in a historical context, with respect to political events, the economic cycle, policy developments in other domains and the broader integration process. There is a general consensus today that the EU should play a role in coordinating research and investment in large infrastructures; this was lacking prior to 1990, even for terrestrial infrastructure projects.

Policy entrepreneurs look to the past for signs of success or failure. They learn by doing and accumulate knowledge, build up expertise and engage with epistemic communities in highly technical areas. As such, if we wish to understand recent advances in projects such as Galileo, then we cannot examine the issue solely by looking at explicit acts of agenda-setting (speeches, communications, proposals or events) for satellite navigation, but need to grasp the broader nature of the policy-making environment at a given time. We need to acknowledge institutional dynamics in the EU in addition to the integration goals of the Community and Member States at the time. Temporal processes are embedded in the EU institutions. The different sequencing of events can produce different outcomes (Bulmer 2009; Goetz 2014; Howlett and Goetz 2014). As such, a comparative historical approach is a useful complement to traditional integration theory – that is, a single policy area cannot always be understood by analysing it in isolation.

Galileo in the Trans-European Transport Network

The arrival of Galileo on the EU policy agenda should be understood as a corollary to a decade of activity promoting terrestrial infrastructures and 'completing' the Single Market – by no means complete by '1992'. In March 1998 the Council called on the Commission to present recommendations on the European approach to global satellite navigation, which it did with a communication entitled 'Galileo: Involving Europe in a New Generation of Satellite Navigation Services' (Commission 1999). The subsequent inception study (PriceWaterhouseCoopers 2001: 2) to support the development of a business plan for Galileo (see also PriceWaterhouse-Coopers 2003) asserted that the rationale for Galileo 'promoted by the European Commission and European Space Agency' covered the following areas.

- Strategic: to protect European economies from dependency on the systems of other states that could deny access to civilian users at any time and to enhance safety and reliability. The only services currently available are the US global positioning system and the equivalent Russian system, both military, but made available to civilian users.
- Commercial: although Galileo will not be able to charge for the use of its basic service because it is accepted that users need to have free open access, it could become a commercially viable business by providing value-added services that will establish a position in the market alongside the global positioning system.
- Economic: Galileo is expected to secure an increased share for Europe in the equipment market and related technologies, deliver efficiency savings for industry, create social benefits through cheaper transport, reduced congestion and less pollution, and to stimulate employment.

Galileo was perceived from the outset as a commercial venture that would enhance the Single Market, with its likely operating services including: an 'open access service', free to all users and providing basic positioning navigation; 'timing signals' as a new universal service; commercial services based on additional encrypted data, allowing a charge to be made; 'safety of life services', which would provide greater accuracy and integrity of the signal, allowing the user to know within a few seconds if the positioning information had become corrupted; a 'search and rescue service', which identified a user's location to civilian emergency services; and a 'public regulated service' based on a robust signal, resistant to interference or jamming and restricted to certain public security organisations such as the police and fire services. As a result, the Galileo Operating Company would generate revenue from the royalties on chipset sales, paid by equipment providers who incorporated a Galileo chip in their products to allow users to obtain the open access service, and income from service providers who wanted to use the specialised encrypted signals to offer other services (PriceWaterhouseCoopers 2001: 4).

Framing the benefits of Galileo

The cost–benefit analysis carried out at the time made a case for particular benefits in the transport sector, particularly for traffic management across the TEN-T network. It stated that 'the benefits principally arise from air traffic control, marine navigation, and route guidance for motor vehicles. The largest and most robust are generated from the aviation and maritime industries' (PriceWaterhouseCoopers 2001: 8). The analysis claimed that the economics of Galileo did not support investment by the private sector on purely financial criteria, but that the benefit to the European economy would be 'significant'. It acknowledged the existence of market imperfection:

> Many of the benefits, such as improved efficiency in the use of airline fleets, are likely to accrue to consumers rather than be captured by the industries that use Galileo services, because competition will ensure that the value cannot realised in higher prices. Industrial users of the service will not therefore be able to increase margins to make a payment to the Galileo operator. There should however be a case for the public sector to promote Galileo, if it can do so at a cost which represents value for money for the economy as a whole taking account of the wider economic and strategic benefits.
>
> (PricewaterhouseCoopers 2001: 8–9)

The Stockholm European Council agreed to the launch of Galileo in March 2001, but acknowledged that its deployment would depend on mobilising the private sector. In Article 42 of its conclusions it stated:

> In conformity with the Cologne and Nice conclusions, the private sector is required to take up the challenge with regard to participation in and financing of the project through a binding commitment for the deployment phase. The European Council notes that the private sector is ready to supplement the public budgets for the development phase. The European Council invites the Council to define the arrangements necessary for launching the next phase of the project, including establishing a single and efficient management

structure before the end of 2001, be it a joint undertaking under Article 171 of the Treaty, an agency or any other suitable body.

(Stockholm European Council 2001)

Ultimately, the private sector was not as prepared to step in as the European Council had foreseen (for subsequent developments on the public–private partnership and private financing, see Chapter 13). Part Four of the European Commission's crucial 2001 White Paper looking ahead to 2010 was entitled 'Managing the globalisation of transport' (Commission 2001). This report asserted that 'enlargement changes the name of the game' and 'the enlarged Europe must be more assertive on the world stage'. As such, Europe needed three things: a single voice in international bodies, to urgently expand the external dimension of air transport and Galileo – a global programme:

> Technology is meeting with increasing success, and new applications are constantly being discovered. Their market and uses cover a whole range of public and private activities and already include transport (location and measurement of the speed of vehicles, insurance, etc.), telecommunications (network integration signals, bank interconnections, electricity grid connections), medicine (e.g. telemedicine), law enforcement (e.g. electronic tagging), the customs service (field investigations, etc.) and agriculture (geographical information systems). It is therefore clearly a strategically important technology and likely to generate considerable profits.
>
> (Commission 2001: 94)

The Commission framed Galileo in terms of security, referring to the threat of depending on US and Russian systems set up for military purposes, which could be blocked or jammed, as had occurred with the US system during the Kosovo war. However, it also highlighted the practical benefits for transport:

> It [the European Union] will thus have at its disposal a tool essential to its transport development policy. For instance, it will be possible using Galileo instantly to trace goods carried on the railway network, facilitating the development of a just-in-time policy. Galileo will permit highly accurate positioning of ships carrying dangerous cargoes and give the maritime authorities the means to ensure safe navigation, particularly in areas of high traffic density such as the Ushant TSS. The emergency, search and rescue and civil protection services are other applications for which Galileo will offer reliable, guaranteed solutions to the strictest standards.
>
> (Commission 2001: 95)

Looking to previous TEN priority project investments, the Commission articulated the cost of Galileo in a way that was easy for policy-makers to understand, while also stressing its low cost relative to terrestrial rail investments that had already been agreed. Galileo would 'open up access to a potential market of

EUR 9 billion a year in return for an investment equivalent to approximately 150 km of high-speed railway track.' It would also 'revolutionise transport, much as the liberalisation of air transport did before it by creating a niche for low-cost airlines which opened up new markets for tourism; or mobile telephony, which has radically changed people's daily lives' (Commission 2001: 95).

In 2002, the Galileo Joint Undertaking was set up (Council 2002), soon followed by a Regulation on its deployment and commercial operating phases. Following the recommendations of the Van Miert TEN-T high-level group in 2003, the Commission agreed to enlarge the existing list of TEN-T priority projects from 14 to 30, putting back the original timetable for completion from 2010 to 2030. It incorporated Galileo as part of the TEN-T Trans-European Positioning and Navigation Network.

Framing trans-European transport networks

How did Galileo become part of the Trans-European Transport Network? Why was Galileo considered part of transport policy? And what was the provision made for new infrastructure as Europe sought to create the Single Market? This section establishes the background to '1992' and examines in parallel the Commission's moves to push for the development of a Trans-European Transport Network consisting of new high-speed rail infrastructures.

The Single Market and large-scale infrastructures

The 1985 White Paper on completing the internal market provided the impetus for more effective Community planning, acknowledging that:

> [t]he right to provide transport services freely throughout the Community is an important part of the Common Transport Policy set out in the Treaty ... the development of a free market in this sector would have considerable economic consequences for industry and trade.
>
> (Commission 1985, paragraph 108)

It highlighted the incomplete nature of the internal market and identified physical and technical barriers as major obstacles to integration. Yet the bulk of fiscal measures pertained to cross-border business and the harmonisation of tax rates, rather than the infrastructure itself. The total exclusion of rail from consideration is an indication of the tight grip of national governments. State aid policy, infrastructure investment and technical harmonisation were not considered important – there was tunnel vision, with a focus purely on the liberalisation of services (Ross 1998: 50–51).

However, several factors gave impetus to the push for a common transport infrastructure. First, the 1985 decision of the European Court of Justice in the case brought by the European Parliament against the Council for failure to act in the field of common transport policy (Case 13/83 1985) underlined the need to make rapid progress in this area. The European Parliament sought a declaration that the Council had infringed the EEC Treaty by failing to adopt a Common Transport Policy. The Council objected, claiming that the European Parliament lacked competence to bring an action under Article 232 (ex 175). The EP had capacity to bring an action under Article 232 and had observed the conditions of that provision in bringing the action. The action was upheld in part, but rejected where the obligation was too vague to be enforceable. This decision gave the Commission a window of opportunity to forge ahead. Ross (1998: 38) claims that the commitment to coordinated transport policy was a case of playing 'catch-up'. The Commission 'rediscovered' rail and became 'newly intent on assuming a much more activist and multi-faceted role in [high-speed train] HST planning, development and funding' (Ross 1994: 193). It announced a medium-term transport infrastructure programme (Commission 1986a) that encompassed all modes of transport, including a high-speed rail network (Commission 1986b). The European Parliament and its Committee on Transport and Tourism endorsed high-speed rail, branding it of 'decisive importance for the revival of passenger and goods transport by rail' (European Parliament 1987). It proposed an action programme for transport infrastructure, advocating an integrated transport market in 1992 (Commission 1988a).[1]

Second, the Single European Act, which entered into force in July 1987, sought the complete implementation of the internal market by 31 December 1992, aiming to ease the free movement of the Treaty's original four factors: capital, persons, goods and services. The preamble alluded to the need:

> to improve the economic and social situation by extending common policies and pursuing new objectives, and to ensure a smoother functioning of the Communities by enabling the institutions to exercise their powers under conditions most in keeping with Community interests.
>
> (*Official Journal* 1987: L 169, 29. 6. 87)

Third, the 1988 report by Paolo Cecchini (Commission 1988b) assessed the advantages of removing these obstacles, quantifying predictions of a one-off net increase in total European Community growth of 5 per cent. Although this prediction was naïve in its over-optimistic assumption of full implementation, it formed an intellectual foundation for political debate; transport-related problems were established as one of the 'costs of non-Europe'.

Fourth, in her famous speech to the College of Europe in Bruges, British Prime Minister Margaret Thatcher asserted that 'by getting rid of barriers, by making it possible for companies to operate on a European-wide scale, we can best compete' with other markets (Thatcher 1988). She added that 'in air transport, we have taken the lead in liberalisation and seen the benefits in cheaper air fares and wider choice' and 'of course we must make it easier for goods to pass through frontiers. Of course we must make it easier for our people to travel through the Community'. In his federalist-oriented address, *A Necessary Union,* to the College one year later in 1989, Jacques Delor referred to the need for thinking big and being far-sighted, while taking account of worldwide geopolitical and economic trends, the movement of ideas and the development of the fundamental values (Delor 1989). In essence, pragmatism, ambition and long-term vision were recognised as essential ingredients for Community policy-making.

The Commission's expert study group, Group Transport 2000 Plus, put infrastructure on the agenda in 1989 and 1990 at the Strasbourg and Dublin Councils. It recognised the existence of a 'direct threat' to the Community's main objectives of cohesion. It identified overcrowding, time-wasting, pollution, inefficiency, a lack of 'inter-modalities' and under-investment, in addition to mode-specific problems such as air traffic control (Ross 1998: 21). In February 1990, Belgian Transport Commissioner Karel van Miert convened an independent study group to advise on infrastructure: European rail was 'in a state of crisis' (Ross 1994: 194). The Commission convened its High-Level Group on the Development of a European High-Speed Train Network. The Group invited national administrations, the Community of European Railways, major rail manufacturers, the Eurotunnel Group and the European Roundtable of Industrialists to send representatives. The three sub-groups set up to define network priorities, investigate technical compatibility and devise an integrated control system held 40 meetings in 1990 alone (Commission 1991a).

The 1990 trans-European networks action plan and priority projects

A decade before Galileo, the Commission proposed an action plan for TENs (Commission 1990a) in the fields of transport, energy and telecommunications. It reflected a renewed recognition of the relationship between effective infrastructure quality and economic development and made the case for the creation of a cohesive series of networks that would 'weave' the continent into a single cloth, bringing transport infrastructure to the forefront of the political agenda. As Johnson and Turner (1997: xi) argued, such projects were consistent with the development of network economies and supported the Commission's principal concerns of maximising the benefits of the Single Market, competitiveness, industrial policy, social and economic cohesion, and the development of closer and ever more interdependent links with Eastern Europe and the Mediterranean. TENs became relevant, not only in 'the great game of integration', but were vital as 'a practical response to the everyday functioning of ... increasingly internationalised economies' (Johnson and Turner 1997: xii).

Developing rapidly within European policy-making between 1990 and 1993, TENs were included within Article 129b-d of the Treaty on European Union (TEU), which reaffirmed the European Commission's objective to 'facilitate the free movement of persons, while ensuring the safety and security of its peoples' (Johnson and Turner 1997: 26–27). The Declaration of European Interest (Commission 1992) that emerged in 1992 paved the way for the Community financing of projects of EU-wide importance. The old concept of harmonisation was 'resurrected' as a Community priority, this time with 'ambitious spending plans to back up planning efforts' (Ross 1998: 51).

Given the inclusion of telecoms and energy as part of the TENs, these networks stood in their own section within Title XII, notably distinct from Title IV on Transport. The TEU gave the EU three principal tasks: to lay down guidelines that identified projects of common interest; to support projects via financial feasibility studies and to provide loans or interest rate subsidies; and to take measures, including technical standardisation, to ensure the interoperability of networks. The TEU even called to extend TEN projects and pushed to accelerate the development of strategic railway lines, roads, airports and waterways. Article 129b of the TEU represented a considerable coup for the sectoral Directorate-Generals to institutionalise policy, supporting the argument that trade policy alone could not bring about all the benefits promised by the internal market (Shaw and More 1995: 213). TENs were needed in areas of transport, telecoms and energy to achieve open and competitive markets and promote interconnection and interoperability.

Trans-European networks were arguably a meta-policy, given their application to regional policy, the environment, cohesion, competition, employment and enlargement. Ambitious, large-scale infrastructural investment on a continental scale offered a 'one policy fits all' political response to short-term economic ills and more long-term structural problems. However, the various actors and institutions saw TENs differently: as a catalyst for growth; as a tool for reducing centre–periphery disparities; and as an instrument for laying the foundations for the accession of the Central and Eastern European countries. As such, we can identify several 'stories' in the policy rhetoric – these narratives or stories contained 'frames' purporting to the aims and benefits of transport infrastructures as a policy alternative.

Frames at play in trans-European transport networks

Frame set 1: growth, competitiveness and employment

Following Maastricht, the Commission reinforced its agenda-setting role. Its 1993 strategic programme to reinforce the effectiveness of the internal market (Commission 1993a) identified the importance of infrastructure in exploiting the benefits of the Single Market; it sought harmonisation and interoperability and promoted the development of networks as crucial to the functioning of the market (Johnson and Turner 1997: 7). The Commission regarded 1992 as 'an important turning point in the evolution of the CTP ... towards a more comprehensive policy designed to ensure the proper functioning of the Community's transport systems' (Commission 1993b).

The 1993 White Paper *Growth, Competitiveness and Employment* called for the speeding up of the implementation of TENs projects. Infrastructural initiatives were seen as essential to achieve significant cuts in unemployment by 2000 (Commission 1993c) and to overcome the inherent imperfections in the market, which had led to a sub-optimal allocation of resources within the EU's 'transaction economy' (Tsoukalis 1993: 80). As the European Commissioner for

Transport made clear, the problem was physical: road congestion was costing Europe an annual 2 per cent of GDP, or nearly €120 billion (Kinnock 1995). Rail offered freight transport at a crawling average speed of 16 km/h and its share of the freight market had decreased from 32 to 15 per cent since the early 1970s (Kinnock 1998). The Commission stated '…the absence of open and competitive markets is hampering, to differing degrees, the optimum use of existing networks and their completion in the interests both of consumers and operators' (Commission 1993c: 76).

The White Paper identified a requisite 'double integration' of national networks and transport modes, towards the creation of a single TEN, by improving existing infrastructures, using more energy-efficient modes of transport to bring about interconnection and modernisation, and promoting technical research (Commission 1993c). The 'conquering' of missing links was assumed to promote a physical integration by connecting national networks, but also political integration through the convergence of philosophical and ideological stances vis-à-vis the approach to cross-border transport infrastructure planning.

The White Paper referred to 'bottlenecks and missing links in the infrastructure fabric' and 'a lack of interoperability between modes and systems' that was hampering access to markets and the workplace (Commission 1993c: 75). The Commission's aim was to create 15 million new jobs and halve the unemployment rate by 2000. Conservative predictions of job creation from TENs, when fully implemented, were of an increase in Community GDP in excess of €500,000 million by 2030 and creating between 600,000 and 1,000,000 new permanent jobs, although how these figures were to be reached was not demonstrated. The Commission stated:

> Networks are the arteries of the Single Market. They are the lifeblood of competitiveness, and their malfunction is reflected in lost opportunities to create new markets and hence in a level of job creation that fall short of our potential.
>
> (Commission 1993c: 75)

Frame set 2: social, economic and territorial cohesion

That which, in the Treaty of Rome, was termed 'harmonious development' became increasingly referred to as 'economic and social cohesion' in the Single European Act (Wallace and Wallace 1996: 210). As a redistributive policy at Community level, TENs could also bolster the construction industry. Investing in physical communications was therefore a remedial policy for European development (Commission 1991b). As Johnson and Turner (1997: 17) pointed out, the philosophy of economic integration embodied in the Treaties suggests that structural interventions were originally only designed for the short to medium term and that in the long term the operations of the market would eradicate regional disparities.

In its first cohesion report, the Commission argued that social and economic welfare could not be guaranteed by free market dynamics. Thus TENs was framed as a cohesive tool intended to combat the competitive forces unleashed by the completion of the Single Market and to prevent them from 'overwhelm[ing] the weaker parts of the Union' (Commission 1996a). By 'improving the spatial balance of the Union', investments would tackle the threat faced by peripheral regions, which risked failing to secure the same gains from integration as more developed and centrally located regions.

> Whether in Ireland, with natural and pressing concerns about peripherality, or in Belgium with its growing problems of congestion, it is equally clear that businesses using transport have a common interest in achieving high quality, modern transport links across the whole continent.
>
> (Kinnock 1996)

There was a clear conflict between the idea of competition to improve the workings of the market and the use of structural funds to 'interfere' in the working of this market (Wallace and Wallace 1996: 211). TENs could be used to stimulate economic growth by lowering the 'friction' of transport between peripheral and central areas. Economic and social gains could potentially be offset by the otherwise negative effects of infrastructural developments that afflict underdeveloped regions, without contributing to peripheral prosperity (Johnson and Turner 1997: 34–35).

> The real challenge is to ensure smooth interplay between, on the one hand, the requirements of the single European market and free competition in terms of free movement, economic performance and dynamism and, on the other, the general interest objectives.
>
> (Commission 1996a, paragraph 19)

Whereas the Single European Act sought to aid economic liberalisation, TENs were framed as a way of intervening in the market by harmonising operating conditions and thus enhancing the potential for economic integration and cohesion (George 1996: 277; Johnson and Turner 1997: 23). In turn, greater cohesion and more balanced levels of growth would lay the foundations for economic and monetary union and facilitate further steps towards political integration. However, growth in the internal market was predicted to induce a greater demand for transport use, encourage modal shifts and lead to an increased pressure for non-polluting ('clean') transport (Johnson and Turner 1997: 16–44).

The need to secure not just a harmonised, but also a harmonious, operating network saw institutional responses to develop common sustainable transport policy. The Commission began to increase the emphasis on environmental impact assessments as a decision-making tool when assessing project proposals, which, in addition, required approval by Council and the European Parliament. Moreover, the Commission recognised that satellite positioning and timing offered 'opportunities for applications in many domains from navigation to surveying, agriculture, oil and gas exploration and others' (Commission 1998: 1).

As satellite navigation became more accepted for use in increasingly precise applications, there would be opportunities to rationalise existing infrastructure. As such, it felt the Communities should bolster the relationship between the terrestrial and satellite components of the Trans-European Positioning and Navigation Network (Commission 1998: 2). The Commission's focus was thus on ensuring the better, more efficient management of existing infrastructures, an objective that Galileo would help achieve. As the Commissioner for Transport asserted: 'just as important is the job of making existing infrastructure work better for the purpose of enhancing sustainable mobility' (Kinnock 1997).

Frame set 3: European interest, additionality and 'network effects'

The Commission was under pressure to justify the added value of policy formulated in Brussels. It needed to establish institutional legitimacy, as well as European 'additionality'. TENs were thus framed as a way to 'bring added value to the *acquis* in acting upon the factors for dynamism in the market' (Commission 1993b; Johnson and Turner 1997: 22). TENs and the internal market were complementary policy areas. As the Community action plan indicated, Europe would have to accommodate anticipated increases in intra-EU trade ('volume effect'), be physically compatible ('interoperability requirement'), offer a European outlook to infrastructure development ('dimension effect'), assure the service standards across the EU ('quality requirement') and promote European economic development ('cohesion effect') (Commission 1990a). Investment would enhance the gains from trade creation relative to trade diversion, reducing the costs of interaction. By influencing the geographical spread of trade patterns, it might bring lower operational costs for business (Johnson and Turner 1997: 16). Community spending on transport infrastructure through the Cohesion and Structural Funds could be coordinated with funding decisions made at Member State level.

The assertion of Johnson and Turner (1997) that 'there was never going be a truly transnational network, only a series of heterogeneous national subsystems' therefore suggests greater justification for Community planning to meet the demands of contemporary economic and political interdependence. Common planning could 'exploit the network metaphor to draw together various strands of infrastructure policy' (McGowan 1993: 183–184). Deriving holistic gains from a set of individual projects depended on connections with other parts of the network built in neighbouring regions. Yet a 'network effect' could only be derived if solid partnerships and a consensus were established among Member States as to the common priorities. Commission-sponsored research indicated 'the benefits from a single project can thus only be realised in full if other projects, often in other regions, are undertaken. This "super-additivity" effect makes it difficult to ascribe precise costs and benefits' (COST 328 1998:19).

The decision-making process for each investment was subject to 'the operation of the key principles governing the general Community approach: concentration, programming, partnership, additionality and value' (Commission 1996b:

2). Politically, the TENs network plan was 'an illustrated expression of common support for the Commission to take charge of infrastructure planning by creating a framework of opportunity that was both additional and complementary to the national one' (Commission 1996b: 2). The need to demonstrate 'additionality' gave further credibility to the Commission's claim that money was better spent if pooled to the Community purse (Bache 1998:126) before being redistributed to cross-border transport projects – each euro of Community spending providing a greater overall return than each euro of national public expenditure (Commission 1996b: 3). TENs would channel resources based on geographical concentration, mobilising resources for cross-border links in the central and peripheral regions. By focusing on large-scale projects, the policy avoided fragmentation to offer high profile and highly visible outcomes of Community policy. Large-scale infrastructure investment would secure political support from the 'poor four' (then Spain, Portugal, Ireland and Greece).

The Commission proposed a Declaration of European Interest (Commission 1992), acknowledging the need to demonstrate 'unequivocal political support for projects subject to the rules set out in Title XII of the TEU'. Projects had to generate direct economic effects in the Community. TENs were framed as an example of how the Single Market could 'protect and improve the satisfactory provision of general services to the public on the basis of universal service or public service obligations' (Commission 1996a). The Commission sought to liberalise services, examining how to combine the dynamism of open markets with general interest requirements. It asserted that 'the Treaty contains means of supporting the European model of society in several ways, for example as a backup for general interest roles: TENs, Community research, consumer policy and social and economic cohesion' (Commission 1996a: 15). Coordinated transport planning and investment would reduce tensions across Member States given the benefits for social welfare and teleological goals of enhanced integration.

Conclusion

The activism of the Commission in TENs is clear. It shaped European infrastructure planning by formulating policy that went beyond the individual interests of Member States, effectively pooling their priority projects. Policy-making in the 1990s increasingly meant a coordination of national and European economic and political interests, even if ultimately the Commission 'pooled' national priorities to create a EU 'wish list' for investment. Political activity was greatest thanks to the renewed confidence of the Commission's Directorates-General, including Transport (DGVII), Energy (DGXVII) and Telecommunications (DGXIII). The leading dynamism of the Flemish Commissioner for Transport Karel van Miert (1989–1993) also helped DGVII (Transport) assert itself vis-à-vis the Member States.

The Commission was entrepreneurial from the mid-1980s and increasingly successful in getting initiatives adopted by the Council. It used its informational resources to derive frames and create favourable conditions for TENs and then

Galileo. It was effective in securing decisions on TENs priority projects, ensuring that each Member State had at least one project of 'Community interest'; TENs might be criticised as an intergovernmental 'shopping list'. The new TEN-T Fund complemented the Cohesion Fund (1992) as a means to help finance projects. Private–public partnerships were also explored, but with limited success given the risk and unattractiveness of the long-term financial returns on investment; the European Investment Bank stepped in. Importantly, despite political agreements at European summits, there was ultimately no legal obligation placed on Member States (or penalties for failing to deliver) regarding TENs. This would affect the Commission's subsequent ability to secure timely implementation, but may also explain why TENs could proceed rapidly – framing a set of individual national investments as a European 'network', while emphasising the completion of 'missing links' were the key to this implementation.

Common policy-making in the field of terrestrial transport infrastructures created a springboard for Galileo (Commission 1999). Just as TENs were 'discursively coupled' with the Single Market project so that the physical completion of the Single Market became a policy solution and wider political goal, Galileo was also framed as helping to secure market integration. Arguably, this is an example of functional spill-over. Completing the Single Market remains a work in progress. Liberalising services and ensuring the free movement of goods and people continues to rest on securing better traffic management and the efficient use of infrastructures – benefits that Galileo and European space policy are 'framed' as helping to deliver.

Note

1 It set out seven projects designated as being of European interest: (1) the Paris–Brussels–Cologne–Amsterdam–London high-speed link; (2) the Lisbon–Seville–Madrid–Barcelona–Lyon high-speed rail link; (3) the modernisation of the Alpine–Brenner axis; (4) the improvement of European air traffic control; (5) the modernisation of the Ireland–north Wales–England road axis; (6) the completion of Scanlink; and (7) the reinforcement of land links in Greece. The Council approved a slightly altered list that included the trans-Pyrenean road link (Somport), but discarded air traffic control, which would re-emerge within the mandate of the Commissioner for Transport and Energy, Loyola de Palacio (1999–2004).

References

Bache, I. (1998) *The Politics of European Union Regional Policy: Multi-Level Governance or Flexible Gatekeeping?* Contemporary European Studies Series. Sheffield: UACES/Sheffield Academic Press.

Bulmer, S. (2009) '*Politics in Time* meets the politics of time: historical institutionalism and the EU timescape'. *Journal of European Public Policy*, 16(2): 307–324.

Case 13/83, *European Parliament* v. *Council of the European Communities* (*Parliament* v. *Council*) (1985) ECR 1513. Available from: www.eur-lex.europa.eu by searching judgement no. 61983J0013.

Commission (1985) *White Paper: Completing the Internal Market.* COM(85) 310 final, 14 June 1985.

Commission (1986a) *Medium-term Transport Infrastructure Programme. Communication from the Commission to the Council.* COM(86) 340 final, 27 June 1986.

Commission (1986b) *Towards a High-Speed Rail Network.* COM (86) 341. 30 June 1986.

Commission (1988a) *Proposal for an Action Programme in the Field of Transport Infrastructure with a View to the Completion of an Integrated Transport Market.* COM(88) 340 final, 16 June 1988.

Commission (1988b) *Europe 1992: The Overall Challenge (summary of the Cecchini report).* SEC(88) 524 final, 13 April 1988.

Commission (1990) *Towards Trans-European Networks for a Community Action Programme. Communication from the Commission to the Council and the European Parliament.* COM(90) 585 final, 10 December 1990.

Commission (1991a) *Proposal for a Council Decision Concerning the Establishing of a Network of High Speed Trains. Communication from the Commission to the Council Regarding a European High Speed Train Network. Sec(90) 2404 final, 6 February 1991.* Available from: www.aei.pitt.edu/5695/1/5695.pdf. Last accessed 15 May 2015.

Commission (1991b) *Europe 2000. Outlook for the Development of the Community's Territory.* COM(91) 452 final, 7 November 1991.

Commission (1992) *Proposal for a Council Regulation Introducing a Declaration of European Interest.* COM(92) 15 final, 24 February 1992.

Commission (1993a) *Reinforcing the Effectiveness of the Internal Market. Working Document.* COM(93) 256 final, 2 June 1993.

Commission (1993b) *The Future Development of the Common Transport Policy: A Global Approach to the Construction of a Community Framework for Sustainable Mobility. White Paper.* COM(92) 494 final, 2 December 1992.

Commission (1993c) *Growth, Competitiveness, Employment: the Challenges and Ways Forward into the 21st Century – White Paper. Parts A and B.* COM(93) 700 final/A and B, 5 December 1993.

Commission (1996a) *Services of General Interest in Europe. Communication from the Commission.* COM(96) 443 final, 11 September 1996.

Commission (1996b) *First Report on Economic and Social Cohesion 1996.* Available from: www.aei.pitt.edu/42144/. Last accessed 11 May 2015.

Commission (1996c) *A Strategy for Revitalising the Community's Railways. White Paper.* COM(96) 421 final, 30 July 1996.

Commission (1998) *Towards a Trans-European Positioning and Navigation Network Including a European Strategy for Global Satellite Navigation Systems (GNSS). Communication from the Commission to the Council and European Parliament.* COM(1988) final. 21 January 1998. Available from: www.aei.pitt.edu/6801/1/6801.pdf. Last accessed 15 May 2015.

Commission (1999) *Galileo. Involving Europe in a New Generation of Satellite Navigation Services. Communication from the Commission.* COM(99) 54 final, 10 February 1999.

Commission (2001) *European Transport Policy for 2010: Time to Decide.* COM(2001) 370, 12 September 2001.

COST 328 (1998) *Integrated Strategic Transport Infrastructure Networks in Europe, 1995–1998. Cordis Report.* Available from: www.cordis.europa.eu/cost-transport/src/cost-328.htm. Last accessed 11 May 2015.

Council (2002) *Council Regulation (EC) No. 876/2002 of 21 May 2002: Setting up the Galileo Joint Undertaking.* Available from: www.eur-lex.europa.eu/legal-content/EN/TXT/?uri=CELEX:32002R0876. Last accessed 11 May 2015.

Delor, J. (1989) *A Necessary Union, Bruges, 17 October 1989.* Available from: www.cvce.eu/content/publication/2002/12/19/5bbb1452-92c7-474b-a7cf-a2d281898295/publishable_en.pdf. Last accessed 15 May 2015.

European Parliament (1987) *Starita Report. Resolution on a High-Speed Rail Network.* Resolution A/279/87, 27 May 1987. See: www.europa.eu/rapid/press-release_IP-89-22_en.htm?locale=en.

European Parliament and Council (1996) *Decision No. 1692/96/EC of 23 July 1996 on Community Guidelines for the Development of the Trans-European Transport Network.* Available, with amendments, from: www.ec.europa.eu/transport/wcm/infrastructure/grants/2008_06_20/2007_tent_t_guidlines_en.pdf. Last accessed 15 May 2015.

European Parliament and Council (2004) *Decision No. 884/2004/EC (added to the list by Decision No. 884/2004/EC), Amending Decision No. 1692/96/EC on Community Guidelines for the Development of the Trans-European Transport Network.* Available from: www.ec.europa.eu/ten/transport/legislation/doc/2004_0884_en.pdf. Last accessed 15 May 2015.

George, S. (1996) *An Awkward Partner: Britain in the European Community,* 2nd edn. Oxford: Oxford University Press.

Goetz, K. (2014) 'Time and power in the European Commission'. *International Review of Administrative Sciences,* 80(3): 577–596.

Howlett, M. and Goetz, K. (2014) 'Introduction: time, temporality and timescapes in administration and policy'. *International Review of Administrative Sciences,* 80(3): 477–492.

Johnson, D. and Turner, C. (1997) *The Political Economy of Integrating Europe's Infrastructure.* Basingstoke: Palgrave Macmillan.

Kinnock, N. (1995) *Three EU Transport Challenges for the Nineties and Beyond: TENs, Environment and Safety.* Speech, Transport Seminar, Rosenbad, Sweden, 29 June 1995. Available from: www.europa.eu/rapid/press-release_SPEECH-95-134_en.htm. Last accessed 11 May 2015.

Kinnock, N. (1996) *The Transport Challenge – European Perspectives.* Address to the Belgian Business Association, Onrad Hotel, Brussels, 24 September 1996. Available from: www.europa.eu/rapid/press-release_SPEECH-96-222_en.htm. Last accessed 11 May 2015.

Kinnock, N. (1997) *Investing in Sustainable Mobility.* Brussels, 21 January 1997. Available from: www.europa.eu/rapid/press-release_SPEECH-97-12_en.htm. Last accessed 11 May 2015.

Kinnock, N. (1998) *Trans-European Rail Freight Freeways.* The Economist Conferences, Hotel Sheraton, Brussels, 16 January 1998. Available from: www.europa.eu/rapid/press-release_SPEECH-98-3_en.htm. Last accessed 11 May 2015.

McGowan, F. (1993) *Air Transport.* European Economy/Social Europe Reports and Studies No. 3: Market Services and European Integration. Brussels: European Commission.

PricewaterhouseCoopers (2001) *Inception Study to Support the Development of a Business Plan for the Galileo Programme. Prepared at the Special Request of DG TREN.* Available from: www.ec.europa.eu/dgs/energy_transport/galileo/doc/gal_exec_summ_final_report_v1_7.pdf. Last accessed 11 May 2015.

PricewaterhouseCoopers (2003) *Galileo Phase II Executive Summary*. Available from: www.ec.europa.eu/transport/facts-fundings/evaluations/doc/2003_galileo_phase_2.pdf. Last accessed 11 May 2015.

Ross, J. (1994) 'High-speed rail: catalyst for European integration?' *Journal of Common Market Studies*, 32(2): 192–214.

Ross, J. (1998) *Linking Europe – Transport Policies and Politics in the European Union*. Santa Barbara, CA: Praeger.

Shaw, J. and More, G. (eds) (1995) *The New Legal Dynamics of European Union*. Oxford: Clarendon Press.

Stockholm European Council (2001) Available from: www.europa.eu/rapid/press-release_PRES-01-900_en.htm.

Thatcher, M. (1988). *Speech to the College of Europe ('The Bruges Speech'), 20 September 1988*. Available from: www.margaretthatcher.org/document/107332. Last accessed 15 May 2015.

Tsoukalis, L. (1993) *The New European Economy Revisited*. Oxford: Oxford University Press.

Wallace, H. and Wallace, W. (1996) *Policy-making in the European Union*. Oxford: Oxford University Press.

10 Framing or hegemony?

The European Metalworkers' Federation and EU armaments and space policies

Iraklis Oikonomou

Introduction

The chapter provides a detailed analysis of the discursive patterns articulated by the European Metalworkers' Federation (EMF) in the interrelated policy fields of space and defence during the period 2002–2004. The main empirical finding is that labour, in the form of the EMF, offered its wholehearted support to every single move that the EU institutions undertook with the arms and space industry to strengthen the latter's competitive status. The chapter argues that the discursive regularities and the maintenance of a stable, pro-industrial stance at the core of labour's rhetoric cannot be explained satisfactorily by frame analysis. Such a set of patterns is instead the outcome of the exercise of hegemony by the ruling social force of the internationalised aerospace industry, through the mediation of ideology and the incorporation of labour into a web of interconnections that falls under the label of an EU politico-military industrial complex.

Why the European Metalworkers' Federation? First of all, because it was an active, recognised and legitimate stakeholder in EU space and defence policies, while also comprising the largest representation of organised labour (the unionised working class) in the field of aerospace at the European level, representing 72 affiliated organisations, four associate member organisations and 5.5 million workers (EMF 2007). It was an organisation with a nature that essentially expanded into the realm of organised interests, social groups and class fractions. Therefore we might expect to see the multiplicity of interests, discourses and narratives that form part of 'framing theory' flourish in this particular institutional environment. The timeline of the chapter focuses on the first half of the 2000s, when the early traces of a truly coherent EU armaments and space policy can be located, with the emergence of the European Defence Agency (EDA) and a series of military-related space projects. However, the findings are consistent with the latter period up until the end of the EMF's life and its transformation into a new institution, industriALL, in 2012, as shown by subsequent documentation.

Methodologically, this chapter is a case study of the role of social forces in European defence and space policy integration. It initially sets a critique of an established theory and provides an alternative analytical framework based on the core premises of historical materialist theory. It then studies key moments of dis-cursive practices as illustrations of a particular argument, before returning to theory and incorporating the findings into the paradigm of the military–industrial complex. Epistemologically, the chapter belongs to the paradigm of critical social science, for which social reality exists independently of the ability of the agency to perceive it. At the same time, reality is mediated by ideational factors that inform action and reflect class and other social structures. In other words, the production and reproduction of knowledge, ideology, discourse and col-lective consciousness are constitutive parts of the social totality and need to be taken seriously by social science (Gill 1993: 28).

In terms of structure, the chapter starts with a theoretical section that presents a rough critique to framing theory and the core principles of a neo-Gramscian alternative on the basis of the notion of class hegemony and the materiality of ideas. The following section documents empirically the stunningly constant demonstration of consensus secured via the EMF concerning a range of meas-ures that unfolded in the formative years of EU space and armaments policy. The material comprises formal declarations, press releases and policy documents produced by the Federation, which are used by the study as illustrations of a spe-cific pattern: the ideological subjugation of labour to the interests and visions of the internationalised military–industrial capital, Thereafter, the analysis turns to the position of the EMF and labour in the EU politico-military–industrial complex, the prime institutional context for the maintenance of hegemony at the EU level. A short conclusion sums up the empirical and theoretical findings of the chapter.

From framing to hegemony: the materiality of ideas

The notion of framing appeared in European integration studies as a reaction to the dominance of a supposedly a-social mainstream consensus, grounded on an understanding of interests as objective givens. Framing has, primarily under the aegis of new institutionalism, fed into several analyses of a broad spectrum of EU policies, such as EU biotechnology policy (Daviter 2012) and EU defence and armaments policies (Mörth 2000). More recently, Stephenson (2012) inserted the concept in the field of EU space policy, drawing on the case of the Galileo satellite project. The expansion of the use of framing as an analytical tool in European politics has been an episode in the broader dominance of idea-tional approaches in the theory of European integration. It may be indeed argued that the notion of 'framing' has yielded fruitful results in terms of both theoret-ical and empirical output, shifting the emphasis from a supposed objectivity of interests to the linkages between interests, ideas and institutions that are inher-ently subjective. Nevertheless, there is a critical problem with this approach, namely, the lack of a theory of power embedded in it or, in other words, the ontological primacy of discourse and ideas over material interests.

The concept of framing is central to sociological institutionalist analysis, where frames 'are referents for action and give direction to the integration process ... legitimate certain decisions and activate certain issues, actors and types of knowledge ... concern power – the power to define and conceptualise' (Mörth 2000: 173–174). In the case of EU armaments cooperation, Mörth (2000) argued that there exists a frame competition among different sections of the Commission, involving two diverging views: the market frame and the defence frame. Armaments cooperation involves an interaction between various actors at different policy-making levels that form networks and informal authority relations and strive to spread their ideas and promote ideational change that later leads to institutional and policy change (Mörth and Britz 2004: 967–971).

The main problem with this theory is that it detaches institutions and organisations from their social environment and conceals their particular functions as channels and mediators for the promotion of the interests of the ruling social forces. In addition, it overlooks the broader politico-economic context in which an institution operates. This choice can only produce a partial and limited explanation of an institution's purpose or intent, given that every institution is not a closed system of power, but one that is embedded within a set of power relations outside this institution. However, the primary problem with institutionalist theory is that it takes for granted the very thing that it needs to explain: cooperation. This preoccupation with the study of decision-making and decision-makers, as defined by policy-makers themselves, and the 'conflation of the study of European integration with the institutional practices defined by policy-makers', is rightly viewed by Hazel Smith as an 'institutional fallacy' and an instance of 'institutional bias' (Smith 2002: 265–266).

The analysis of sociological institutionalism in EU defence policy should start where it ends: from the acceptance that 'various parts of the Commission are part of a defence-industrial network' (Mörth 2000: 186). Are the sources of this network only institutional and frame-related ones? Moreover, what are the sources of framing itself? Mörth (2000) hesitantly accepts that frames depend on interests as well as ideas, but she fails to discuss which interests prevail and why, and how specific ideas relate to specific interests. In other words, she does not provide a convincing account of the reasons why different frames tend to coincide and contribute to the same policy objectives. Her analysis describes the emergence of contending frames, but omits the factors that lead to the prevalence of one specific frame over another. For Mörth, 'organizations structurally reflect socially constructed reality' (Mörth 2000: 177). This constructivist claim ignores the fact that reality is also material, subjected to the forces of social production. This materiality generates the social construction and thus the latter should not be conceived in the purely ideational terms of frame competition. Focusing on a hypothetical dominance of particular frames misses the material origins of these frames and the power relations that produce and reproduce them. The discussion of frames should thus lead to the discussion of class interests, as institutions are a complex reflection of social relations and a means of their reproduction. Social relations form the socio-economic environment in which institutions operate and, in this sense, institutions have imprinted in their function the preferences and contradictions of the hegemonic social forces. If class relations condition the function of the institution, then the study of the role of institutions should be the study of the social relations that underpin them. The non-materialist ontology of sociological institutionalism and its lack of a theory relating state interests to class power represent an insurmountable obstacle for such an endeavour.

Dominant ideologies are not given, but produced through political processes that reflect the given configuration of socio-economic power within an existing social order, be it national or transnational. Such processes also involve the participation of actors whose main function is the production of hegemonic discourse and ideology. By no means does this imply the existence of the 'conspiracy' in favour of the European arms industry that has spread within the European research community. What it does imply is that the hegemony of European military–industrial capital necessitates the build-up of sources of legitimacy, some of which can be found within the area of academic institutions and think-tanks, and labour. According to Gramsci (Hoare & Nowell Smith) critical, organic intellectuals accompany the emergence of a dominant class in the realm of social production, giving this class 'homogeneity and an awareness of its own function not only in the economic but also in the social and political fields'. For the purpose of this chapter, the organic intellectuals may be individual or collective and include think-tanks, EU scholars and any other individual or institution that produces and reproduces the hegemonic ideology that legitimises the interests of the EU military–industrial historical bloc. Members of the corporate community can occupy a position of organic intellectual, in accordance with the view of Gramsci (Hoare & Nowell Smith 1971: 5–6) of capitalists as organisers of a broader general interest, beyond their sectoral–corporate interest.

Andreas Bieler and the neo-Gramscian school of international relations have fruitfully attempted to connect ideas with the economic realm, relating ideational production to material production. Consider his core statement: 'ideas represent an independent force, but only in so far as they are rooted in the economic sphere' (Bieler 2000: 16). Ideas have a material basis, given that they are produced in a given context of social relations of power and reflect the preferences and practices of those forces (van Apeldoorn 2001: 73). However, there is no linear link between material reality and ideas, as material forces are contradictory and produce multiple possibilities. Gramsci reiterated Marx's claim that 'it is on the level of ideologies that men become conscious of conflicts in the world of the economy' (Hoare & Nowell Smith 1971: 162).

Bieler (2000) defines ideas as 'inter-subjective meanings' that are constitutive of social practices, reflecting the interrelatedness of agency and structure. In addition, ideas may function as legitimising tools for certain policies, mediated particularly by organic intellectuals – that is, intellectuals that are 'the true representatives of a particular social group, generated by the sphere of production' (Bieler 2000: 13, 16). Ideas both condition the ways individual and collective actors understand their interests and may be used as instruments for the achievement of collective objectives. However, the neo-Gramscian claim over the constitutive function of ideas should not be conflated with the constructivist claim over the constitutive primacy of ideas. Ideas inform interests, but are themselves shaped by the prevailing relations of production and each other's position vis-à-vis these relations; existing social relations of production are then the prime source of the formation of actor's ideas and identities. Moreover, ideas sustain and legitimise class hegemony, a rule that, although rooted in the economic sphere, encompasses political and ideological elements as well.

Hegemony refers 'to a relation between social classes, in which one class fraction or class grouping takes a leading role through gaining the *active consent* of other classes and groups' (Gill 1990: 42). Consent does not exclude coercion; they are both instances of hegemony, sustaining structures of class domination through institutionalised ideas and practices (van Apeldoorn 2002: 20). For Gramsci, hegemony is born out of the process of social production. In his words (Hoare & Nowell Smith 1971: 161), 'though hegemony is ethical-political, it must also be economic, must necessarily be based on the decisive function exercised by the leading group in the decisive nucleus of economic activity'. That said, hegemony does not derive automatically, but is instead a political process, involving a multitude of actors and the mediation of ideology. Hegemony is sustained through political means, even though economic predominance is the decisive element of its emergence.

Role of the European Metalworkers' Federation in EU space and defence policy

The previous section argued theoretically that subordinate classes play a key role in legitimising politico-economic initiatives and allowing the rule of the hegemonic classes by consent. In the context of EU space and armaments policies, organised European labour – in the form of the EMF – allied closely with the military–industrial capital in generating a broader EU policy reform consensus, towards a direction of more military capabilities, more funding for arms productions and the military use of space, more support for the arms and aerospace industries, and the further development of a common European security and defence policy. In other words, labour was co-opted in re-producing the discourse and the strategy of the industry towards a direction of militarisation.

The EMF was founded in 1971, succeeding the European Committee of Metalworkers' Unions that was set up in 1963 by seven founding trade unions from Germany, France, Italy, Belgium, the Netherlands and Luxemburg. The prime areas of activities in recent years included industrial policy and employment, social dialogue, collective bargaining policies and employee representation at the company level. The Federation operated as a distinct organisation up until May 2012, when it was dissolved to join the industriALL European Trade Union. The two other organisations that became part of the new trade union were the European Mine, Chemical and Energy Workers Federation, and the European Trade Union Federation – Textiles Clothing and Leather. Key reasons for this merger were the pre-existing cooperation among the three organisations, the need to cut administrative costs, the achievement of greater representativeness, and the integration of mining and metalworkers' unions under a single umbrella (EMF 2011a: 32). The present study is only concerned with EMF's positions and policy as an autonomous institution before the establishment of industriALL.

The analysis of the EMF's positions reveals the tremendous consensus offered by the organisation vis-à-vis EU armaments and space-related initiatives. In fact, it is impossible to distinguish between the logic of EU initiatives and the logic of EMF; the latter seems to be simply encouraging and welcoming the former. This logic can be split into several themes, or instances, the most obvious being: EU independence from the USA; the development of military capabilities; the establishment of a European Armaments Agency; the protection of the European arms industry's competitiveness; and the exploitation of space as a security tool on behalf of the EU.

European autonomy relative to the USA is at the core of the legitimising of militarisation expressed by the EMF. Speaking about space applications and satellites, the EMF stresses that 'the EU's dependency on U.S. and Russian military systems is of concern' (EMF 2002: 4). There is more to space and competition with the USA:

> it is necessary to show far greater political willingness in certain areas such as aerospace, where much still needs to be done and where Europe is beginning to lag considerably behind compared to the US, i.e. in the area of manned space flights.
>
> (EMF 2002: 12)

The fear of dependency on the USA expands into the realm of defence: 'A strong industrial base is vital to support the development of a European defence and security policy and retain Europe's specialised technological expertise and know-how rather than be reduced to relying on off-the-shelf purchases from America' (EMF 2002: 3–4). Data are mobilised to confirm this argumentation: military aerospace equipment expenditure in the USA, at 20 per cent of the total government defence budget, is higher than in the EU at 13 per cent (EMF 2002: 13).

The narrative on capabilities is linked to the tasks and commitments that the EU is meant to carry out through its common defence policy to tackle security threats and challenges. The then Secretary of the EMF stated: 'We have to give Europe the real ways and means to fulfil its obligations in a world of increasing insecurity and threats to peace' (Kuhlmann, in EMF 2003b: 1). The primary vehicle for such fulfilment is, needless to say, the Common Security and Defence Policy of the EU. As early as 2003, the trade union was calling 'for a renewed impetus to the development of a European common security and defence policy' (EMF 2003a: 2). The conceptual chain 'security – Common Security and Defence Policy – capabilities development' has the European arms industry at its final ring. The call for capabilities development is, in other words, a call for developing capabilities made in Europe, by European industries, rather than capabilities in general. As for the spectrum of what needs to be produced, it seems endless: the EMF repeatedly invited EU institutions to consider 'utilising European sources of technology and production, in order to maintain European capability for total systems development, including platforms, engines and equipment' (EMF 2002: 3).

As for the process that led to the EDA, the EMF was rapid in highlighting 'the importance of the discussions in the European Convention regarding the establishment of a common European armament policy and the creation of a European Armament Agency' (EMF 2003a: 2). One year earlier, the EMF had called for 'the establishment of a European armaments agency with a specific procedure to give a European focus to complete the task of defence equipment acquisition and support' (EMF 2002: 4). In fact, the vision of the EMF was truly ambitious and far-reaching, demanding that:

> the prerogatives attributed to OCCAR [Organisation for Joint Armament Cooperation] be reinforced within the framework of a future European armaments agency in order that a joint body for defence procurement may provide the necessary impetus for the forward-looking development of the defence industry.
>
> (EMF 2004: 9)

Again, one is left wondering why an organisation fighting for workers' rights and employment had to have such a long-term and detailed opinion over what should have been done at the level of military policy, strategy, institutions and decision-making.

Notably, the trade union repeatedly expressed an opinion, not only about the need to support the industry, but also about what form of ownership the industry should have. The answer to the question 'public or private' is clearly in favour of the latter and appears in discursive occasions that seem unrelated to that question:

> The EU wants a military arm and the Commission has decided to promote scientific military research. Previously, it was up to the member states and we welcome this as a step in the right direction. Europeanisation requires privatisation of the arms industry and an energetic pursuit of this goal.
>
> (Peter Shaaf in EMF 2003c: 74)

In the name of competition, privatisation is advertised as a useful tool to liberalise the market: 'To prevent an unacceptable distortion of competition by public and semi-public companies in Europe, ownership structures must be chosen that best guarantee the companies' future in the defence industry in the long run' (EMF 2004). It is understandable that an industrialists' association would push for measures favouring the strengthening of their competitive position. Why this should be a matter of concern for workers, however, is a wholly different matter.

In fact, when it comes to industrial competitiveness, it seems as if the EMF has been talking on behalf of the industry: 'Governments must give the companies clear prospects and reliable starting positions based on a clear industrial policy concept' (EMF 2004: 5). The organisation was one of the first to embrace the findings of the STAR 21 report, calling it 'an important contribution through its approach of improving both the competitiveness and employment of the European aerospace industry' (EMF 2003a: 1). The existence, survival and competitive expansion of European arms manufacturers was pictured as a matter of life and death, in a conjuncture that was then dominated by the drafting of an EU constitution: 'If we want to avoid the failure of the EU constitution, we must ensure we have a solid European defence industry' (Reinhard Kuhlmann in EMF 2003c: 4). The establishment of a European defence equipment market was deemed necessary for the fulfilment of such a goal and the EMF openly demanded

> the creation of equal conditions of competition for all suppliers as part of a common European market. This necessitates in a next step regulations for the opening of national markets as well as the further development of Article 296 of the EU Treaty.
>
> (EMF 2004: 9)

Finally, organised labour has been unquestionably in support of EU military space policy in general and of the Global Monitoring for Environment and Security and Galileo in particular. In its repeated calls for an enhanced set of EU military space programmes, the EMF seems to go beyond even the most ambitious EU strategist:

> Fulfilling Europe's security needs will require the development of suitable programmes. These could range from earth observation to telecommunications with some extra non-space programs. A comprehensive approach would suggest a network centric system. Such a type of program includes all observation means, satellites (infra red, visible and radar), high altitude reconnaissance planes, drones, telecommunications (necessary by satellite for the projection of *our* forces), global positioning and guiding, etc. but should also include tactical means, i.e. more precise and sophisticated systems suitable for crisis management (tactical observation satellites, tactical launchers and tactical launch pads).
>
> EMF (2005: 5)

The ideological submission of the EMF to the rhetoric of capabilities and competition with the USA is indeed striking to the extent that it resembles closely the discourse reproduced by the space industry. The invocation by the EMF of the current USA/EU ratio of space expenditure in favour of the USA, the need for additional budgetary commitments and the emphasis on space research and development all point to this direction. Moreover, elements of protectionism are traceable in the condemnation of solutions that are not Europe-only and involve industrial participation from outside the EU (EMF 2004: 1; 2005: 6). In this context, Global Monitoring for Environment and Security was naturally seen by the EMF (2005: 5) as 'a good initiative', while the 2004 European Commission's White Paper on space policy was welcomed with a reminder of the role that space can play in fulfilling the EU's security and defence goals (EMF 2004).

Labour as a segment of an EU military–industrial complex

The concept of the military–industrial complex has a long history as an established, but also contested, term. For some, the term is scientifically weak. According to Joana and Smith (2006: 72), 'the concept of a "military–industrial complex" has too frequently been used from a normative and polemical angle that is inconsistent with objective social science'. For others it stands as a useful framework, primarily in the context of US civil–military–industrial relations (Lens 1970; Smith and Smith 1983). Smith and Smith (1983: 41) referred to the military–industrial complex as a 'confluence of interests between arms manufacturers and the military establishment'. According to Berghahn (1981; 87), the term denotes a 'coalition between the armed forces and the arms manufacturers for the purpose, not of waging war, but of maintaining large military expenditures'. Focusing solely on the relations between the power of the military industrial capital and the armed forces reflects a normative concern with a supposed control of government by the military and industrialists. Political power is missing from this definition, because presumably military and economic interests have substituted it.

However, the multiple layers of authority and interests and the multiplicity of actors involved in the institutionalised setting of the complex go far beyond a simple industrial–military conjunction. State authority is crucial. For Faramazyan (1978: 13), the military–industrial complex represents 'an alliance of military–industrial monopolies, the military and the state bureaucracy'. Gansler (1980: 72–82) studied the involvement of the Department of Defense, Congress and other political agencies in the functioning of the US military–industrial complex. Lens (1970: 145) defined the complex as 'a conglomerate of elites – a military elite, an industrial elite, a banking elite, a labour elite, an academic elite – which seeks its own aggrandizement through global expansion'. The concept encompasses national and supranational political power, senior military personnel, ideological and discursive practices, academic and expert-based legitimising functions and traditional industrial lobbying practices. Senghaas (quoted in Bergahn 1981: 87) spoke of the existence of a 'political–ideological–military–scientific–technological–industrial complex'.

Even though the roots of a military–industrial complex are to be found in the realm of production, and political power and institutions reflect and reproduce the dominant military–industrial interests, the essence of such a concept is not simply instrumental. In other words, the term not only involves a dense interaction between policy actors, but also contains a solid ideological dimension.

Mészáros (2003) pointed out that 'the expression "military–industrial complex" ... clearly indicates that what we are concerned with is something much more firmly grounded and tenacious than some direct political/military determinations (and manipulations) which could be in principle reversed at that level'. The military–industrial complex denotes shared frames of thinking, shared notions of national security and profound institutional linkages that go beyond temporary bureaucratic alignments of a tactical nature (Gill and Law 1988: 115).

In the EU context, Manners (2006) has used the term 'military–industrial simplex' to define:

> the way in which both the military-armaments lobby and the technology-industrial lobby have worked at the EU level to create a simple but compelling relationship between the need for forces capable of 'robust intervention', the technological and industrial benefits of defence and aerospace research, and the ... creation of the European Defence Agency.
>
> Manners (2006: 193)

Slijper (2005) has superbly documented the influence that the European arms manufacturers exercised in Brussels during the formative period of EU armaments policy, detailing the multi-faceted lobbying activities of the industry and highlighting industrial dominance in expert groups that legitimised the armaments agenda at the time. He rightly predicted that 'As Brussels places more importance on military matters, the arms industry will find cause to intensify its efforts to frame policy, while "levelling the playing field" in ways that are beneficial to their business interests' (Slijper 2005: 34). Finally, Hayes (2006) delivered an impressive account of the making of the Commission's European Security Research Program, conceptually based on the notion of a security–industrial complex.

However, the idea of an EU military–industrial complex is not devoid of problems and challenges. As Lovering (1998: 235) rightly points out, primary weapons contractors are still nationally based, transnational collaborative arrangements involve just a few of the arms-producing countries and many industrial actors are oriented towards non-European countries. It is true that the socio-economic basis of the European arms industry is fragmented, characterised by national differentiations. Above all, we cannot ignore the lack of a single EU super-state. The US military–industrial complex is based on a unified state apparatus that, despite its internal frictions, has a unity of purpose in the support of the US arms industry. The concept of a military–industrial complex has been historically applied to the domestic context, highlighting the ways domestic interests and structures influence decisions in the name of the 'national interest' (Buzan and Herring 1998: 108). What 'national interest' is an EU military–industrial complex supposed to represent and articulate?

The EU military–industrial complex does not substitute respective national complexes, but rather complements them, in the absence of an EU state. It is a web of interconnected actors and mechanisms in favour of internationalised military–industrial interests, operating at the EU level, the initiatives of which run parallel to the function of national defence–industrial establishments. It is not meant to introduce EU defence spending in the place of French or British defence spending – especially as there is no European army and no European budget for military equipment acquisition – but rather to add EU funding to the companies' revenues, next to the money that originates from national defence budgets. In other words, the EU military–industrial complex is not meant to replace, but rather to complement.

The mechanism of communication and consultation between the industry, EU institutions and the labour was formalised in the European Partnership on the Anticipation of Change in the Defence Industry, which was launched in December 2008 and included the Aerospace and Defence Industries Association of Europe, the Commission, the EDA and the EMF. The declared aim of the initiative was to help workers and industrialists 'manage restructuring in a socially responsible way' (European Commission 2008). According to the Federation's positive assessment, 'this partnership promotes activities to foster a culture of jointly anticipating change and developing skills in the sector' (EMF 2011b: 32). The lack of any reference to the political and economic orientation of the other stakeholders is stunning. Equally stunning is the treatment of 'experts' and 'expertise' by the workers' association and especially of 'experts' not affiliated with the organisation itself. The EMF cooperated regularly with external experts and was keen to work with the industry jointly; the formal rationale was the following: 'should other stakeholders organise partnership-related projects the EMF will seek to be involved to advance the interests of European workers' (EMF 2011b: 32).

A channel through which the EMF secured a spot in the drafting of the policies of the complex was the participation of its officials in experts' reports funded by the European Commission, such as LeaderSHIP 2015, drafted in 2003 by the High Level Advisory Group. The Group comprised European Commissioners, Parliamentarians and industrialists, chaired by the then Commissioner for Enterprise Erkki Liikanen. The then General Secretary of the EMF was a full member of the panel. The Group expressed its support for the creation of what was later to become the EDA and requested the setting up of a joint procurement agency to promote consolidation of the sector (European Commission 2003: 25–27). In other cases, the EMF received direct funding from the European Commission for studies that were awarded to external experts, such as the study on the industrial prospects of land armament systems, which was undertaken by a Spanish research firm and involved consultation with the Commission and the EDA (EMF 2012: 21). The Commission also funded public events organised by the Federation: 'Thank you to the EC for the subsidies that were allocated to us to organise this conference. The Commission has not influenced at all the content of this conference' (Kuhlmann in EMF 2003c).

Far from being linear and smooth, this process of labour co-optation is replete with contradictions. Access to EU institutions is not uniform. The EMF had no major formal or informal links to the EDA, but maintained good working relations with the Commission and DG Enterprise (personal communication, EMF official, Brussels, 28 March 2007). In addition, contrasting national legacies and the respective socio-economic positions of capital had an impact on the coherence of labour intervention. For example, the EMF Working Group on defence remained idle after the beginning of 2006 due to fractional disagreements on procurement, ownership and restructuring policies. A Nordic fraction, including Germany, sponsored a private-oriented restructuring, while a southern fraction under France sought to protect the public and semi-public ownership status of their industries (personal communication, EMF official, Brussels, 28 March 2007). Such contradictions are to be expected, given the contrasting national legacies and socio-economic backgrounds, and do not alter the big picture discussed in this chapter.

Conclusion

Analysis of the EMF's public discourse reveals how it fully embraced the agenda of EU armaments policy in the critical years of the early 2000s. In fact, the EMF publicly invited European governments to harmonise military requirements, establish a European armaments agency, procure sophisticated military equipment and increase military research and development funding at the EU level in the name of industrial competitiveness and defence capabilities. The EMF extended its advocacy beyond support for military–industrial interests to include calls for the strengthening of a common European security and defence policy. In this context, the military and security applications of space were predictably favoured by the labour organisation in a most consistent manner. To put it simply, the organisation co-authored and co-signed the book of EU militarisation as a collective organic intellectual of European arms and space manufacturers in the name of job protection and workers' interests.

To explain labour's pro-industrial stance, we may argue that this was part of a policy to protect employment in the European arms sector. Trade unions are supposed to fight for employment and a competitive sector is supposed to generate or at least sustain employment. This account is, however, not satisfactory. The EMF advocated consolidation and privatisation instead of – for example, conversion – even though these processes were linked to excessive job losses. At the culmination of EU-wide consolidation in 1999, employment in the arms industry had fallen by more than one-third compared with 1990 (Defence Analysis Institute: 2003: 37). Instead, the positions of labour should be viewed as a politico-ideological outcome, as a moment of the hegemony of the European aerospace and arms industry. The consistency and depth of labour's support for industrial positions and the uncritical adoption of the industry's agenda reflect the power of the ideological representations of the hegemonic social force: the internationalised European military–industrial capital.

Why is labour necessary for the production and reproduction of the hegemony of an EU politico-military–industrial complex? The short answer is 'legitimacy'. The long answer should indicate capital's need to depict its interest as the general interest – that is, as the common sense by which all actors should abide for the progress and prosperity of the entire society – in this case, the EU. The addition of labour to this group of forces is a public and rather vocal reinforcement of legitimacy. This is not the 'bad guys' of the arms industries talking; it is labour, the workers, the people. The full adoption of capital's vision has rendered the elevation of labour to the status of a transnational counter-hegemonic force impossible. The agenda of the conversion of the arms industry from military to civilian markets remains the only viable and progressive alternative to militarisation. However, due partly to the processes described in this chapter, no social force capable of pushing this agenda forward has emerged yet.

References

Apeldoorn, B. van (2001) 'The struggle over European order: transnational class agency in the making of "embedded neo-liberalism"'. In: A. Bieler and A.D. Morton (eds), *Social Forces in the Making of the New Europe: the Restructuring of European Social Relations in the Global Political Economy*. Basingstoke: Palgrave Macmillan, 70–89.

Apeldoorn, B. van (2002) *Transnational Capitalism and the Struggle over European Integration*. London: Routledge.

Berghahn, V. (1981) *Militarism: the History of an International Debate, 1861–1979*. Cambridge: Cambridge University Press.

Bieler, A. (2000) *Globalisation and Enlargement of the European Union: Austrian and Swedish Social Forces in the Struggle Over Membership*. London: Routledge.

Buzan, B. and Herring, E. (1998) *The Arms Dynamic in World Politics*. London: Lynne Rienner.

Daviter, F. (2012) *Framing Biotechnology Policy in the European Union*. ARENA Working Paper 05/2012. Oslo: ARENA Centre for European Studies.

Defence Analysis Institute (2003) *Prospects on the European Defence Industry*. Athens: Defence Analysis Institute.

European Commission (2003) *LeaderSHIP 2015: Defining the Future of the European Shipbuilding and Shiprepair Industry*.

European Commission (2008) 'EU partnership to help 800,000 workers in the defence industry better adapt to change'. *Press Release, IP/08/1898, 8 December 2008*.

European Metalworkers' Federation (2002) *Managing Change in the European Aerospace Industry – Final Draft*. Policy Document, 28 October 2002.

European Metalworkers' Federation (2003a) *Open Up the Skies and Space for the Future – React to the Problems on the Ground Now!* Press Release, FEM 15/2003.

European Metalworkers' Federation (2003b) *Ensuring a Future for the Defence Industry in Europe Lies in Industrial Policy Dialogue with Workers' Representatives*. Press Release, FEM 55/2003, 11 December 2003.

European Metalworkers' Federation (2003c) *The Future of the European Defence Industry*. Conference Proceedings, 10–11 December 2003.

European Metalworkers' Federation (2004a) *White Paper – a European Commission Space Policy Action Plan*. Policy Document, 10 March 2004.

European Metalworkers' Federation (2004b) *Defence Technology in Europe: Political Responsibility, Industrial Development and Employment Policy Perspectives*. Memorandum Adopted by the 97th EMF Executive Committee, 7–8 June 2004.

European Metalworkers' Federation (2005) *The EMF View of What Should Be the European Space Policy*. Policy Paper, 6 June 2005.

European Metalworkers' Federation (2007) *The EMF Today & Tomorrow*. Information Brochure.

European Metalworkers' Federation (2011a) *40 Years 1971–2011, European Metalworkers' Federation*. Trilingual Report on the History of EMF.

European Metalworkers Federation (2011b) *Work Programme 2011–2015 of the European Metalworkers' Federation*. Policy Document Approved by the 4th EMF Congress, 9–10 June 2011.

European Metalworkers' Federation (2012) *Report on Activities 2011–2012*. Policy Report, 15 May 2012.

Faramazyan, R. (1978) *Disarmament and the Economy*. Moscow: Progress.

Gansler, J.S. (1980) *The Defense Industry*. Cambridge, MA: MIT Press.

Gill, S. (1990) *American Hegemony and the Trilateral Commission*. Cambridge: Cambridge University Press.

Gill, S. (1993) 'Epistemology, ontology and the "Italian school"'. In: S. Gill (ed.), *Gramsci, Historical Materialism and International Relations*. Cambridge: Cambridge University Press, 21–48.

Gill, S. and Law, D. (1988) *The Global Political Economy: Perspectives, Problems and Policies*. London: Harvester & Wheatsheaf.

Gramsci, A., Hoare, Q. and Nowell Smith, G. (eds) (1971) *Selections from the Prison Notebooks of Antonio Gramsci*. New York: International.

Hayes, B. (2006) *Arming Big Brother: The EU's Security Research Programme*. TNI Briefing Series No. 2006/1. Amsterdam: Transnational Institute.

Joana, J. and Smith, A. (2006) 'Changing French military procurement policy: the state, industry and "Europe" in the case of the A400M'. *West European Politics*, 29(1): 70–89.

Lens, S. (1970) *The Military–Industrial Complex*. Philadelphia, PA: Pilgrim Press.

Lovering, J. (1998) 'Rebuilding the European defence industry in a competitive world: intergovernmentalism and the leading role played by companies'. In: M. Kaldor, U. Albrecht and G. Scheder (eds), *Restructuring the Global Military Sector: Volume II. The End of Military Fordism*. London: Pinter, 216–238.

Manners, I. (2006) 'Normative power Europe reconsidered: beyond the crossroads'. *Journal of European Public Policy*, 13(2): 182–199.

Mészáros, I. (2003) 'Militarism and the coming wars'. *Monthly Review*, 55(2) (June). Available from: www.monthlyreview.org/2003/06/01/militarism-and-the-coming-wars. Last accessed 29 September 2014.

Mörth, U. (2000) 'Competing frames in the European Commission – the case of the defence industry and equipment issue'. *Journal of European Public Policy*, 7(2): 173–189.

Mörth, U. and Britz, M. (2004) 'European integration as organizing: the case of armaments'. *Journal of Common Market Studies*, 42(5): 957–973.

Slijper, F. (2005) *The Emerging EU Military–Industrial Complex*. TNI Briefing Series No. 2005/1. Amsterdam: Transnational Institute.

Smith, D. and Smith, R. (1983) *The Economics of Militarism*. London: Pluto Press.

Smith, H. (2002) 'The politics of "regulated liberalism": a historical materialist approach to European integration'. In: M. Rupert and H. Smith (eds), *Historical Materialism and Globalization*. London: Routledge, 257–283.

Stephenson, P. (2012) 'Talking space: the European Commission's changing frames in defining Galileo'. *Space Policy*, 28(2): 86–93.

11 The Member States of the European Space Agency

National governance structures, priorities and motivations for engaging in space

Christina Giannopapa, Maarten Adriaensen and Daniel Sagath

Introduction

Understanding the past and current strategy, policy and governance structures of Members States is the key to preparing future strategies and setting up a coherent European space policy and programme. As put forward in this book, framing theory can be used in this process (Kohler-Koch 1997, Mörth 2000; Baumgartner 2001; Sabatier 2011). There are several political loci with different specificities that create and influence the actor's frames. The actors engaging in these loci influence the political process within a certain frame or narrative that has been shaped by their background (March and Olsen 1996; Sabatier 2011). Space is a particularly interesting and dynamic field because it includes many horizontal and vertical frames in which policies emerge.

At the top of the vertical axis, the European Space Agency (ESA), as an intergovernmental organisation, brings actors together under a specific setting and steers a peaceful, scientific, industrial and cooperative frame. The ESA was created in 1975 by 10 founding Members States; there are now 20 Member States and the organisation is still growing. The ESA is Europe's gateway to space, with a mission to shape the development of Europe's space capabilities and to ensure that investments in space continue to deliver benefits to the citizens of Europe and the world. There are a number of articles in the ESA Convention that are relevant to a European space policy and programme, as well as to the Member States successfully participating in this framework. One of the most defining legal provisions that directly effects the ESA's frame is Article II of the ESA Convention, which describes the purpose of the ESA and gives it the power to formulating policies:

> [It can be achieved] by elaborating and implementing a long-term European space policy, by recommending space objectives to the Member States, and by concerting the policies of the Member States with respect to other national and international organisations and institutions; by elaborating and

implementing activities and programmes in the space field; by coordinating the European space programme and national programmes, and by integrating the latter progressively and as completely as possible into the European space programme, in particular as regards the development of applications satellites; by elaborating and implementing the industrial policy appropriate to its programme and by recommending a coherent industrial policy to the Member States.

(ESA 2003: 13–14)

At a lower level of the vertical axis, European Member States shape the policy formulation process at the national/ESA/EU level. The past decade saw an increasing number of accessions to both the ESA (Poland was the twentieth state to accede to the ESA Convention on 19 November 2012) and the EU (Croatia was the twenty-eighth state to enter the EU on 1 July 2013). Figure 11.1 shows the overlap of the ESA and EU Member States and the ESA Cooperating States. The number of European Member States composing a space strategy and policy and defining their space governance for the first time is increasing, while other countries may consider revising their space policies with the current changes in the European space landscape. Even though the ESA has had long-standing success in European cooperation in space activities and policy-making power through its Convention, this cooperation is typically on a programme-by-programme basis. There have been limited systematic efforts towards the exchange and coordination of national space strategies and programmes to create a coherent and holistic European space policy and programme.

The current efforts towards establishing a European space policy is the joint work of the European Commission and the ESA through the eight joint Space Councils. Two EU flagship programmes have emerged from the close cooperation between the EU and ESA: Galileo, which is the EU's navigation system and the observation system Copernicus (formerly Global Monitoring for Environment and Security or GMES) (Giannopapa 2012). With the increase in the number of Member States engaging in space, the dynamic process and outcome will change as new frames emerge and add to the political bargaining process – the important question is how this process will change.

The space sector relies on institutional markets with limited room for global competition. The role of institutional funding defines the structure of the industry and its competitiveness. European industry today performs well in both institutional and commercial markets and has well-known excellence in both the manufacturing and services of telecommunication satellites and launchers. Its competitiveness relies on the stable support of public investment, which is fundamental for financial sustainability.

This chapter aims to map the frames at the national level in Europe and to give a holistic assessment of the national space governance structures, strategic priorities, motivations in engaging in space activities and industrial structure in the 20 ESA Member States. This will help to give a better prediction of future space developments in Europe.

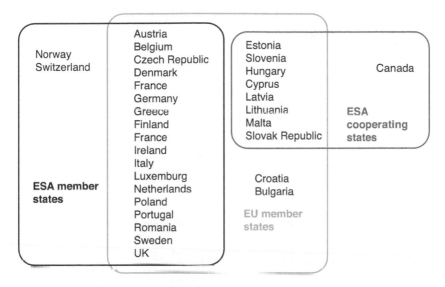

Figure 11.1 Member States and cooperating states of the European Union and the European Space Agency.

European Member States in the ESA

Since its creation, the number of Member States in the ESA has doubled. The ESA is expanding its membership to the countries that have joined the EU since 2004. The Czech Republic, Romania and Poland have all become full members and Hungary and Estonia are soon to start negotiating to become full ESA Member States. Eight EU Member States have signed a Cooperation Agreement with the ESA and two are negotiating such an agreement. Figure 11.2 gives an overview of the ESA integration process (Sagath *et al.* 2013).

A Member State joining the ESA goes through a transition period of between five and nine years. In this period, the following objectives need to be achieved (ESA 2012): balanced participation in the mandatory programme; acceptable overall industrial return; development of a national space strategy providing guidelines, priorities and resources; a formal national structure to support the delegation that forms the interface with the ESA and liaison with industry; envisaged subscription to selected optional programmes; effective national space industrial association; balanced participation of industry and academia; at least one payload, sensor or sub-system to be flown on an ESA mission; at least one company involved in the supply chain of one of the Large System Integrators[1]; and/or at least one sustainable commercial activity related to services or applications exploiting space. Member States that have joined the ESA within the last ten years are considered as New Member States for the purpose of this chapter. Table 11.1 compares the characteristics of the New Member States of the ESA according to these criteria.

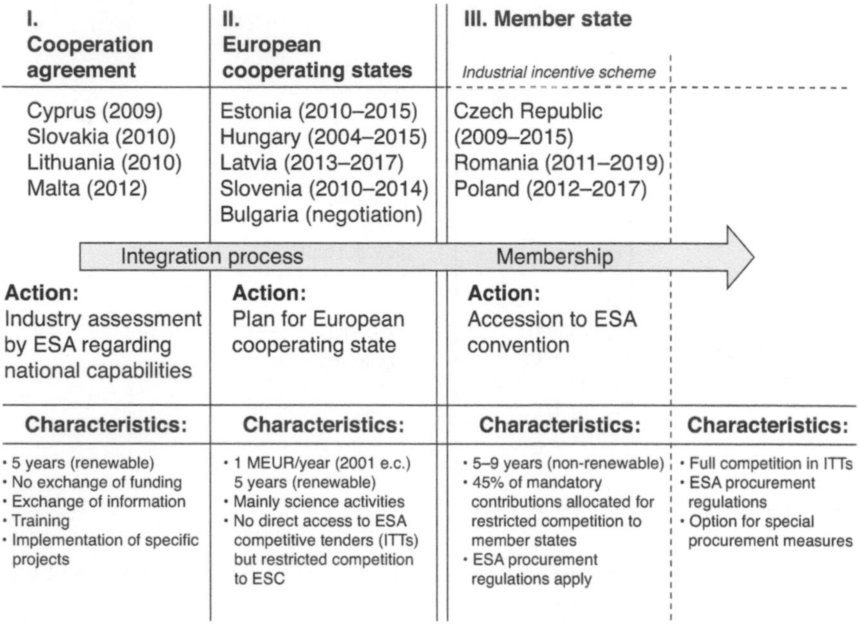

I. Cooperation agreement	II. European cooperating states	III. Member state *Industrial incentive scheme*	
Cyprus (2009) Slovakia (2010) Lithuania (2010) Malta (2012)	Estonia (2010–2015) Hungary (2004–2015) Latvia (2013–2017) Slovenia (2010–2014) Bulgaria (negotiation)	Czech Republic (2009–2015) Romania (2011–2019) Poland (2012–2017)	

Integration process → Membership →

Action: Industry assessment by ESA regarding national capabilities	**Action:** Plan for European cooperating state	**Action:** Accession to ESA convention	
Characteristics:	**Characteristics:**	**Characteristics:**	**Characteristics:**
• 5 years (renewable) • No exchange of funding • Exchange of information • Training • Implementation of specific projects	• 1 MEUR/year (2001 e.c.) 5 years (renewable) • Mainly science activities • No direct access to ESA competitive tenders (ITTs) but restricted competition to ESC	• 5–9 years (non-renewable) • 45% of mandatory contributions allocated for restricted competition to member states • ESA procurement regulations apply	• Full competition in ITTs • ESA procurement regulations • Option for special procurement measures

Figure 11.2 Process of integration into the European Space Agency.

Member State participation in the ESA Mandatory Activities is related to their respective GDP, while participation in the ESA Optional Programmes is by subscription. There is no official grouping of the ESA Member States within the ESA. For the purpose of this chapter, when plotting the contribution to the ESA of the Member States against their GDP, we have placed the Member States into three groups (Figure 11.3). The first group is those Member States with a budget greater than €200 million: Germany, France, Italy and the United Kingdom. The second group includes those with a budget between €50 million and €200 million: the two front-runners, Belgium and Spain, followed by Switzerland, the Netherlands, Sweden, Norway and Austria. The third group is those with a budget between €10 million and €30 million: Poland, Denmark, Finland, Ireland, Romania, Portugal, Luxemburg, Greece and the Czech Republic.

ESA Member States space governance

Each ESA Member State has a unique governance structure when it comes to space. The space governance structure determines: who has the power of decision-making and to what extent; who has the lead in representation; and how integration between stakeholders is made. Frames define and diagnose policies and justify the proper governance and decision-making process (Baumgartner and Mahoney 2008: 436).

Table 11.1 Comparison of the new Member States of the European Space Agency

Country	National strategy	Space office/agency	Industry association	Inter-ministerial co-ordination	ESA space budget
Portugal 2000	2004 to be updated 2013	Space office (FCT)	Yes	Through space office	Jointly from various ministries
Greece 2005	No	Proposal has been made	Yes	No	GSRT only
Czech Republic 2008	2014	In process of establishing an agency	Yes	Inter-Ministerial Coordination Council	Jointly from various ministries
Romania 2011	No	Romanian Space Agency (ROSA)	No	Through ROSA + Inter-ministerial Group for Security and Research	Ministry of Education and Research, Youth and Sports
Poland 2012	2012 Action Plan expected 2013	In process of establishing an agency	Yes	Inter-Ministerial Board for Space Policy	Jointly from various ministries

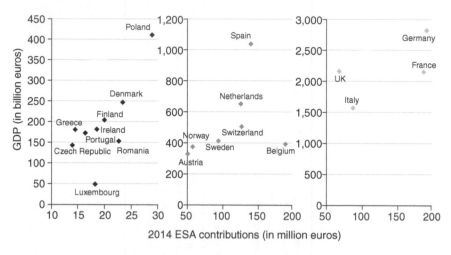

Figure 11.3 GDP versus 2014 contributions of the European Space Agency Member States (source: International Monetary Fund 2014).

Each Member State participates in a number of organisations engaged in space activities and international agreements (Table 11.2). The organisations include the ESA, the European Organisation for Exploitation of Meteorological Satellites (EUMETSAT), the EU Satellite Centre, the European Global Navigation Satellite Systems Agency (GSA), the European Defence Agency (EDA), the International Telecommunications Union (ITU) and the United Nations Committee on the Peaceful Uses of Outer Space (UNCOPUOS). The typical ministries involved in overseeing these organisations are related to: science, technology, research and education; the economy; industry and innovation; transport, communications, defence, environment, energy and foreign affairs; and the Prime Minister's office. Figure 11.4 shows the ministries responsible for overseeing these organisations. This analysis of the governance structure of the 20 ESA Member States was carried out through semi-structured interviews with stakeholders and through official documents.

Different ministries focus their attention on different aspects of space policy in seeking to support their positions (Baumgartner and Mahoney 2008). The national institutes for meteorology, which are often located under the umbrella of the Ministry of Environment, constitutes in many cases the delegation to EUMETSAT. Greece and Italy are exceptions; in these states the responsible ministry for EUMETSAT is the Ministry of Defence, highlighting the importance attached to meteorology for national security and defence purposes. Delegations to the European Defence Agency and EU Satellite Centre are also linked with the Ministry of Defence and representation to UNCOPUOS is linked with the Ministry of Foreign Affairs. Aspects related to satellite communications are often under the competence of the Ministry of Transport and/or Communications.

Table 11.2 European Space Agency Member States subscription to international organisations and ratification of international agreements

Member State	International organisations							International agreements									
	EU	ESA	EUMETSAT	ITU	ESO	EISC	NATO	OST	FA	LC	RC	MA	NPT	MTCR	WA	CTBT	HCOC
Austria	1995	1986	1993	X	X	–	–	X	X	X	X	X	X	X	X	X	X
Belgium	1957	1978	1986	X	X	X	X	X	X	X	X	X	X	X	X	X	X
Czech Republic	2004	2008	2010	X	X	X	X	X	X	X	X	–	X	X	X	X	X
Denmark	1973	1977	1986	X	X	–	X	X	X	X	X	–	X	X	X	X	X
Finland	1995	1995	1986	X	X	–	–	X	X	X	–	–	X	X	X	X	X
France	1957	1980	1986	X	X	X	X	X	X	X	X	–	X	X	X	X	X
Germany	1957	1977	1986	X	X	X	X	X	X	X	X	–	X	X	X	X	X
Greece	1981	2005	1986	X	–	–	X	X	X	X	X	–	X	X	X	X	X
Ireland	1973	1980	1986	X	–	–	–	X	X	X	–	–	X	X	X	X	X
Italy	1957	1978	1986	X	X	X	X	X	X	X	X	–	X	X	X	X	X
Luxembourg	1957	2005	2002	X	–	X	X	X	–	X	–	–	X	X	X	X	X
Netherlands	1957	1979	1986	X	X	–	X	X	X	X	X	X	X	X	X	X	X
Norway	–	1986	1986	X	–	–	X	X	X	X	X	–	X	X	X	X	X
Poland	2004	2012	2009	X	–	–X	X	X	X	X	X	–	X	X	X	X	X
Portugal	1986	2000	1986	X	X	–	X	X	X	–	–	–	X	X	X	X	X
Romania	2007	2011	2010	X	–	X	X	X	X	X	–	–	X	–	X	X	X
Spain	1986	1979	1986	X	X	X	X	X	X	X	X	–	X	X	X	X	X
Sweden	1995	1976	1986	X	X	–	–	X	X	X	X	–	X	X	X	X	X
Switzerland	–	1976	1986	X	X	–	–	X	X	X	X	–	X	X	X	X	X
UK	1973	1978	1986	X	X	X	X	X	X	X	X	–	X	X	X	X	X

Acronyms (see United Nations 2014 for details of international agreements):
EUMETSAT, European Organisation for Exploitation of Meteorological Satellites; ITU, International Telecommunications Union; EISC, European Interparliamentary Space Conference; NATO, North Atlantic Treaty Organization; OST, 1957 Outer Space Treaty; RA, 1968 Rescue Agreement; LC, 1972 Liability Convention; RC, 1975 Registration Convention; MA, 1978 Moon Agreement; NPT, 1968 Non-Proliferation Treaty; MTCR, 1987 Missile Technology Control Regime; WA, 1995 Wassenaar Arrangement; CTBT, 1996 Comprehensive Nuclear Test Ban Treaty; HCOC, 2002 Hague Code of Conduct against Ballistic Missile Proliferation.

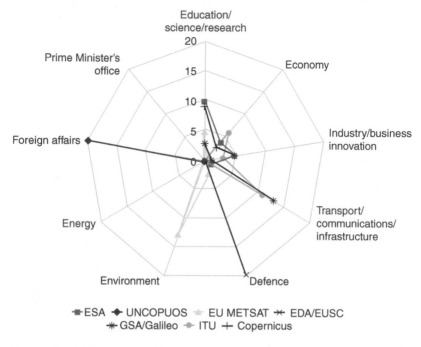

Education/
science/research

Economy

Prime Minister's
office

Foreign affairs

Industry/business
innovation

Transport/
communications/
infrastructure

Energy

Environment

Defence

➡ ESA ◆ UNCOPUOS ▲ EU METSAT ✳ EDA/EUSC
✳ GSA/Galileo ● ITU ✛ Copernicus

Figure 11.4 Ministry responsible for space in European Space Agency Member States.

This is reflected by the implementation of policy and regulations by telecommunications agencies and the representation of the states by these agencies in international organisations dealing with telecommunications, such as the International Telecommunications Union. Ministerial delegations to the EU dealing with Galileo and European Global Navigation Satellite Systems Agency are typically linked with the national Ministries for Transport and Telecommunications, regardless of which ministry is responsible for the ESA.

In the case of the ESA, in 10 Member States the overseeing ministry is in charge of education, science and/or research; in five Member States ESA affairs resort under the Ministry of Economy; in four the Ministry of Industry and/or Innovation is responsible; and in one Member State it is the Ministry of Transport that is in charge. In the case of France, space has traditionally been administered under the Ministry of Higher Education and Research (at least for civil space activities). In Germany, the Netherlands and Finland, the ministries in charge of economic affairs take the lead. In Germany and the Netherlands, in particular, space used to be under the responsibility of the Ministry of Science and Technology, but has now moved to the Ministry of Economy, highlighting the transversal nature of space. In Poland, the responsibility lies with the Ministry of Economy. In Ireland, Norway and the UK, the ministries driving innovation and supporting industry have the prime responsibility for space activities. In the Czech Republic, the competence for ESA affairs has been allocated to the Ministry of Transport. In the case of other countries ESA cooperates with, such as Turkey and Israel, space is under the responsibility of the Prime Minister, highlighting the geopolitical importance of space.

Another trend in ESA space governance is the sharing and/or delegating of responsibility and budgets for space by multiple ministries – for instance, transport and (tele)communications, environment, energy or defence. In practice, one ministry has the leading responsibility for the ESA space activities, with one or multiple other ministries having secondary space responsibilities. Examples are the Czech Republic, Portugal and Denmark. In the Czech Republic, the Ministry of Transport has the lead for space. In the exercise of its competence it is, however, assisted by a wide variety of other players, including: the Ministry of Foreign Affairs; the Ministry of Education, Youth and Sports; the Ministry of Industry and Trade; the Ministry of Environment; the Czech Hydrometeorological Institute; the Ministry of Justice; the Czech Environmental Information Agency; and the Czech Academy of Sciences.

This kind of division of responsibility of space at the national level is managed by the space agency, space office, or as part of a department or unit that also deals with other topics. The Member States with a space agency are Austria, France, Germany, the United Kingdom, Italy and Romania. In Member States where a space agency exists, all space activities at the national, European and international level are handled by the agency, including participation in the UN. However, there are variations in the agencies regarding their powers in setting up a national space strategy, awarding national contracts, conducting research and development, making binding commitments vis-à-vis the ESA; and representation to other organisations. In the Member States with a space office, the authority is in principle smaller than for the space agencies, although, again, their power is determined and governed by the national context. In some cases the space office plays a central role, coordinating and makes links between all the space activities of the country, whereas in other cases higher fragmentation can be found. If no agency and no office is present, the implementing entity is typically a unit or department within a ministry.

Figure 11.5 shows an archetype model for space governance based on patterns observed in the space governance of the ESA Member States. Each state's particular governance is based on its own national environment and specificities. Regardless of the type of governance structure chosen by a Member State, the following need to be achieved: a coordinated approach towards the various fields of space activities; no duplication of effort; setting out a national space policy/strategy and overseeing its progress; coordinating representation in space-related bodies; and coordinating a coherent space budget (Sagath *et al.* 2014).

Overall, various governance models are identified depending on the responsible ministries for space for each of these organisations. Regarding the ESA, traditionally, education or science ministries are in charge. Other states have opted to resort space under the Ministry of Economy or Industry and Innovation. Recognising the transverse nature of space and its potential role for a number of

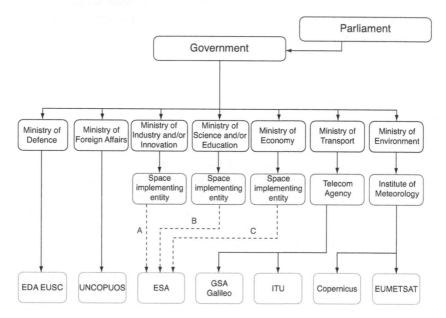

Figure 11.5 Archetypal governance model for space.

sectorial policies, transport or environment can be selected to take the lead in space. The three implementing entity types of agency, office and department or unit are a matter of choice for the Member States. However, regardless of the choice made, it is necessary that they are empowered with a coordinating role for national space activities to avoid incoherence, fragmentation and duplication of efforts.

The ESA frame and current activities reflect the tension and consensus between its actors: the ministries responsible for space and the national space implementing entities. Space policy, both at the EU, the ESA and the Member State level, is continuously in the process of redefining and repositioning itself in the frame of the political discourse in which space resorts. The following section demonstrates how this inherent process is driven by underlying rationales and motivations for Member States to engage in space activities. These reasons, in turn, are based on and evolve with the European and national frame for space activities.

Priorities of ESA Member States

Understanding the varied priorities in space of the Member States is a complex issue. They have been framed to cover three thematic areas: technology; sustainability; and motivation. Framing here can be seen as the process of selecting, emphasising and organising the priorities (Daviter 2007). The strategies of the

Member States were analysed together with information provided at the workshop Exchange of National Strategies and Plans, which took place at the ESA in March 2014.

The priorities with regard to the technology domain vary from one Member State to another (Figure 11.6). Most Member States have a broad interest over the entire scale of technology domains. The relative importance of each technology domain differs between Member States. Nevertheless, it can be observed that all Member States see potential in the development of their downstream space sector; integrated applications are considered by all Member States. Furthermore, science and exploration are highly regarded by all Member States. This may be a result of the foundation of the ESA and the fact that its mandatory programmes all include science. All other technology domains (navigation, satellite communications, launchers and human space flight – including micro-gravity research) are well represented, partially due to historical reasons and industrial interests.

Space, as a result of its transversal nature, can serve as an important 'multiplier' to the six areas of sustainability (Schrogl 2009): security, environment, energy, resources, knowledge and transport. Member States foresee a different level that space can have as an enabling tool in support of these areas (Figure 11.7). Space is considered to be a strong tool in support of various policy areas. These are mostly a logical consequence of the concerned Member States' historical, geopolitical, economic, geographical, financial and political positions and outlooks. For the ESA Member States with long land and/or sea borders, border security is an important area to which space can contribute. In Member States with considerable natural resources (including fisheries, mining and energy), their priorities tend to coincide with serving these interests. It is not surprising that transport and communications are generally accepted as an important area for contributions from space assets (Sagath *et al.* 2014).

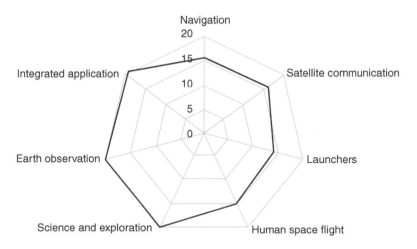

Figure 11.6 Priorities for space in technology domains.

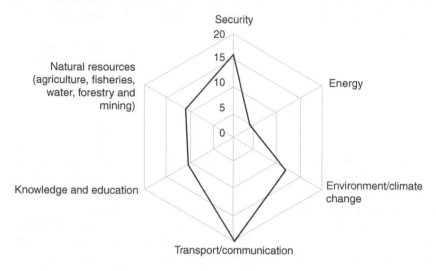

Figure 11.7 Priorities for space in areas of sustainability.

There are a number of identified motivators that encourage the involvement of Member States in space activities: boosting industrial competitiveness; engagement in international cooperation; technology development and transfer; job creation; European non-dependence; and social benefits (Figure 11.8). Through the analysis of these motivators for the ESA Member States, it was found that social benefits as a motivator score relatively low compared with the other motivations that governments use as rationale to place funds and manage space activities. The top motivator for space investments is to increase industrial competitiveness, followed closely by the notion of promoting and fostering international cooperation.

The ESA Member States unanimously perceive investment in space as a means to enhance the competitiveness of their respective space and space-related industries, or high-tech industry in general. Space is without a doubt an ideal area for international cooperation as it is outside of the scope of national territories, a common good and too expensive for one state alone to engage in. Closely linked with the objective of industrial competitiveness, the potential for technology transfer from space for commercial purposes in terrestrial applications is highly valued by the ESA Member States. Job creation follows next, as the sector is perceived as high-tech with a need for highly skilled personnel. The role of space in European non-dependence is also an important rationale for public investment in space. This is typically a stronger motivator for the larger space-engaged Member States, such as France. This need for independence from other global space actors can mainly be found in access to space, core satellite technology in navigation, satellite communications and Earth observation, and independent access to satellite data for decision-making (Sagath *et al.* 2014).

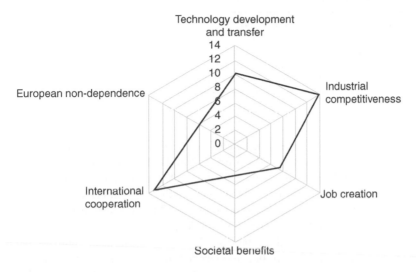

Figure 11.8 Motivations for space.

Industrial structure of the ESA Member States

In 2013, the European space upstream industry had an annual turnover of €6.8 billion and the public institutional market represented 53 per cent of the final turnover. The ESA represents 34 per cent of industry sales and 67 per cent of the total institutional share. This makes the ESA the largest institutional customer in Europe. Thus the ESA plays an important role in shaping the industrial structure landscape of the industries of its Member States.

Concentration ratios are used to show the extent of market control of the largest firms in the industry. The structure–conduct–performance paradigm says that more firms lead to more competition and lower prices. Thus the market structure – that is, how far the structure departs from the perfect competition model – determines the conduction of the firms in terms of pricing and production decisions or inter-action with other firms. This, in turn, determines their performance. Two measures of market concentration used by economists are the N-firm concentration ratio and the Herfindahl–Hirschman index. The N-firm concentration ratio is defined as the sum of the output proportions of the N largest firms in an industry. Typically, the first four and first eight largest firms are used for the calculation of the CR4 and CR8 values. The Herfindahl–Hirschman index is defined as the sum of the squared market shares of all firms in an industry. A high market concentration is repres-ented by a high N-firm ration or a high Herfindahl–Hirschman index. There are different levels set for the interpretation of the Herfindahl–Hirschman index. The one used here is that a Herfindahl–Hirschman index below 1000 indicates an non-concentrated index; between 1000 and 1800 indicates moderate concentration; and above 1800 indicates a higher concentration. Table 11.3 shows the CR4 and CR8

Table 11.3 Concentration ratio of the European space industry

Concentration	2003–2007	2008	2009	2010
CR4 (%)	51	81	82	83
CR8 (%)	80	89	90	91
Herfindahl–Hirschman index	928	3.235,6	3.455,7	3.435,1

Source: ESA 2013.

values and the Herfindahl–Hirschman index of the European space industry and how it has evolved over time. As shown, the Herfindahl–Hirschman index increased by over 200 points between 2008 and 2009, and both the CR4 and CR8 values showed corresponding increases. Concurrently, the number of companies with market shares >0.5 per cent has dropped from 12 to 11, which has also had a negative effect on the Herfindahl–Hirschman index.

Over the past few years, the concentration of the European industry has increased. Large companies have become larger and national interests have continued to play a major role in investment decisions. OHB, which has emerged as a new large system integrator, is an example. Although the emergence of OHB suggests a shift in concentration, the overall tendency points towards the dominance of the prime companies and, in particular, three primes: Airbus, Thales Alenia Space and OHB. They together control about 73 per cent of the total space industry and a few governments hold a combined 25 per cent interest in the industry overall (ESA 2013).

Industry concentration is also evident in the supply chain structure. Primes tend towards verticalisation, meaning that they develop units or acquire firms that together create an end-to-end supply chain. This vertical structure then reduces the contracting opportunities for smaller firms, especially in relation to larger contracts. As a further consequence, there are fewer opportunities for small and medium-sized enterprises, and for firms from under-returned and New Member States, to compete for large system integrator business. These trends may result in reduced competition, redundant capacity across the industry and fewer pressures to improve productivity and drive product innovation (Porter 2008). These negative effects of industry concentration may be especially apparent with regard to the ESA's smaller and under-returned Member States. Companies from these countries report difficulties in accessing contracts or, in other words, they experience high barriers to entry owing to the dominance of very large companies.

As the ESA is the largest public institutional actor, examining the industrial participation of the industries of Member States in ESA tenders will provide an indication of the capabilities and structure of the Member States' industries. In addition, it can be used to show the extent of the market control of the largest firm in ESA public tendering and in each Member State. As the ESA is the largest public tender, it provides information on how ESA decisions can affect the industrial structure in Europe. The highest concentrated industry structure is found in Germany, followed by Luxemburg, Sweden, Romania, Austria,

Denmark, Norway, Switzerland and Poland. In the second category, with a moderate concentrated market, is Finland, the United Kingdom, France, Italy and Spain. The countries with the least concentrated markets are Greece, followed by Belgium, the Netherlands, Ireland and Portugal.

The relative absence of concentration of the space industry in these Member States depends on unique factors in the respective countries. In the case of Belgium, the division of industry in the three regional entities (Flanders, Wallonia and the Brussels Capital Region) has historically led to the development of multiple, but smaller, space industry entities, without one or more prime contractor or large system integrator, despite the relative importance of Belgium as the ESA's fifth contributor. Furthermore, Belgium has a wide spread of activities in the ESA, covering all technology fields. A similar situation is seen in the Netherlands, where there is involvement in most programme areas. The presence of the European Space Research and Technology Centre (ESTEC), the largest ESA establishment, has favoured an evolution towards Dutch companies becoming the main local subcontractors for a wide variety of programme areas. Ireland, Portugal and Greece have a similar lack of concentration based on their choices of strategic programme. In the case of Ireland, this is a result of the government's policy choice to support non-space industries to participate in ESA tenders as they see space technology development and applications mainly as technology transfer to other areas – a combination of universities and small and medium-sized enterprises are present. In the case of Portugal, there is a lack of space tradition.

In the next couple of years, it is expected that industry will consolidate, with a few companies covering the largest part of the contracts. Greece's interest in the activities of the ESA has revolved mainly around the participation of academia rather than large industries, leading to a lack of concentration. In the case of Greece, the landscape of the industrial structure relates to the lack of a space strategy, appropriate governance and shifting in priorities, combined with a lack of participation from larger companies and the traditional involvement of the scientific community in space science. Subscription to the ESA Optional Programmes was often directed to programmes that favoured the tendering of multiple smaller contractors rather than large companies. However, the lack of appropriate funds is expected to shift the distribution of contracts.

The New Member States can be subdivided in Member States with a significant space heritage – for example, Luxembourg and the Czech Republic –and those with virtually no space heritage – for example, Portugal and Greece. The challenge for all the Member States is the same: how to successfully compete with established space companies. The manufacturing industries of the New Member States have the potential to direct existing industrial capabilities into the space market by offering products at a lower price. This is a capital-intensive path that, under the current economic situation, is only possible if the national funding entity is willing to support the efforts of industry over many years. The other path is less intensive in terms of capital expenditure, but no less risky. It corresponds to identifying market opportunities in the application of space technologies, often in combination with other technologies. However, for newcomer companies, and also for some established companies, if their space experience and know-how are limited or recent, it may be difficult to access enough market knowledge to identify promising markets and position themselves accordingly. For both paths, strong support by national agencies and the ESA and EU funding mechanisms is mandatory.

Conclusion

Framing theory has contributed to improving our understanding of the political processes. Different discourses are shaping the policy outcomes and these, in turn, are shaped by a plethora of actors. This chapter has mapped the European space landscape in terms of the ESA, the Member States, institutions, industrial structure and prevailing interests.

Each Member State has a unique governance structure. However, engagement in the ESA can be found under three ministries: science and education, industry and innovation, and economy. It is not uncommon that space competences are moved from one ministry to another because the perception of where space can best serve is changing as the sector matures and shows multiplier effects in other policy areas. Traditionally, ministries for science and/or education have been in charge of space. As the importance of space for industry has grown, there has been a shift to the ministries dealing with industrial competitiveness and innovation. The most recent trend is to place space in the Ministry of Economy, highlighting the importance of space as a multiplier of public funding and its transverse nature to a number of policies. The smaller ESA Member States that have recently joined the EU typically place space under a ministry such as transport. However, since the financial crisis there has been an emerging trend to have multiple ministries involved in funding space activities.

Three types of implementing entities are found for space activities: the space agency, the office and the unit. The choice varies according to the Member States' needs, priorities and government structure. A large Member State will typically have a space agency that is competent in being the authority responsible for elaborating space strategy and planning, managing the space budget and representing in the relevant organisations and forums. The space office is a good alternative for Member States that do not have the resources for an agency, but wish to have a centralised overview and coordination of space activities. The unit typically not only deals with space, but also with other topics. Space competences are typically then spread to a number of ministries and bodies with limited coordination. ESA Member States that have a large contribution to the ESA combine strong support for the ESA with that of their national agencies, whereas medium- and small-sized Member States prefer to channel the bulk of their efforts via the ESA as their agency or divide their resources between national and ESA efforts.

The priorities of Member States have been framed to cover three thematic areas: technology, sustainability and motivation. The technology domains where

Member States focus their strategy are: science and exploration; earth observation; integrated applications; satellite communication; launchers; human space flight; and navigation. All Member States engage in science and exploration, Earth observation and integrated applications. Space policy is treated as an enabling tool and 'multiplier' for a number of policy areas, such as security, environment, energy, resources, knowledge and transport. Most Member States place an emphasis primarily on the role of space in communication, transport and security, with less on energy. The motivation in engaging in space activities also varies. The motivators identified are: boosting industrial competitiveness; engagement in international cooperation; technology development and transfer; job creation; European non-dependence; and societal benefits. Industrial competitiveness and international cooperation are the top motivators, whereas societal benefits are at the bottom.

Strategy and policy, space governance and industrial structure continuously shape each other. The current setting reflects the current and evolving political frame for space priorities at the national and supranational level. Three important trends emerge. First, the number of Member States engaging in space is increasing, with common interests in science, Earth observation and integrated applications. Second, economic and international aspects of space with a focus on competitiveness and growth and international cooperation continue to be the main focus. Third, interest in space to support communication, transport and security are the top areas of interest. The increase in the number of actors and the change in the locus of policy elaboration to a number of ministries and the increased engagement of the Ministry of Economics and Transport will have a strong impact on shaping the future European space policy. The need for a common European perception of space by Member States will also make the decision-making process smoother. However, this might also evoke distribution conflicts among competing actors with similar interests for resources. This will lead to debates and trade-offs in areas where interests are not share by all the Member States, such as human spaceflight and launchers.

Acknowledgements

The authors thank Anastasia Papastefanou for her contribution in analysing the governance, priorities and motivation of the Member States.

Note

1 Large System Integrators are enterprises that have, at the same time, at least one ongoing prime contract for over €200 million and concerning space-related infrastructures, launchers or satellites; in this area the enterprise should have either an annual turnover greater than €200 million or an annual balance sheet total of more than €200 million.

References

Baumgartner, F.R. (2001) 'Political agendas'. In: N. Polsby (ed.), *International Encyclopaedia of Social and Behavioural Sciences: Political Science*. New York, NY: Elsevier.

Baumgartner, F.R. and Mahoney, C. (2008) 'The two faces of framing: individual-level framing and collective issue definition in the European Union'. *European Union Politics*, 9(3): 435–448.

Daviter, F. (2007) 'Policy framing in the European Union'. *Journal of European Public Policy*, 14(4): 654–666.

European Space Agency (2003) *Convention for the Establishment of a European Space Agency and ESA Council Rules of Procedure*, 5th edn. Nordwijk: ESA Publications Division.

European Space Agency (2012) *Approach to Implementation of Decisions with Respect to New Member States*. ESA/IPC(2012)33, rev.1, corr.1, Paris, 27 March 2012 (unpublished).

European Space Agency (2013). *Analysis of the European Space Industry 2008–2010*. 4000102658/10/F/MOS (unpublished).

Giannopapa, C. (2012) 'Securing Galileo's and GMES's place in European policy'. *Space Policy*, 28(4): 270–282.

International Monetary Fund (2014) *World Economic and Financial Surveys. World Economic Database October 2014 Edition* [online]. Available from: www.imf.org/external/pubs/ft/weo/2014/02/weodata/index.aspx. Last accessed 25 October 2014.

Kohler-Koch, B. (1997) 'Organized interests in European integration: the evolution of a new type of governance?' In: H. Wallace and A. Young (eds), *Participation and Policy Making in the European Union*. Oxford: Clarendon Press.

March, J.G. and Olsen, J.P. (1996) 'Institutional perspectives on political institutions'. *Governance*, 9(3): 247–264.

Mörth, U. (2000) 'Competing frames in the European Commission – the case of the defence industry and equipment issue'. *Journal of European Public Policy*, 7(2): 173–189.

Porter, M. (2008) 'The five competitive forces that shape strategy'. *Harvard Business Review*, 86(1): 25 40.

Sabatier, P. (2011) 'The advocacy coalition framework: revisions and relevance for Europe'. *Journal of European Public Policy*, 9(1): 98–130.

Sagath, D., Adriaensen, M. and Giannopapa, C. (2013) 'Space activities of the central and eastern European countries: past and present'. Paper presented at the 64th International Astronautical Congress, Beijing, China, 23–27 September, 2013, paper IAC-13-E3.1,2,x20172.

Sagath, D., Adriaensen, M., Giannopapa, C. and Papastefanou, A. (2014) 'An assessment of space governance structures and space strategies in member states of the European Space Agency (ESA)'. 65th International Astronautical Congress, Toronto, Canada, 29 Sept–3 October 2014, paper IAC-14, E3,1.1x23510.

Schrogl, K.-U., Mathieu, C. and Lukaszczyk, A. (eds) (2009) *Threats, Risk and Sustainability – Answers by Space*. Vienna/NewYork: Springer.

United Nations (2014) *Status of International Agreements Relating to Activities in Outer Space as at 1 January 2014* [online]. A/AC.150/CRP.7, Vienna, 20 March 2014. Available from: www.oosa.unvienna.org/pdf/limited/c2/AC105_C2_2014_CRP07E.pdf. Last accessed 25 October 2014.

12 Space as a field and tool of international relations

Veronica La Regina

Introduction

Through the joint influence of the EU and the Member States of the European Space Agency (ESA), Europe has made a number of achievements in the space field (Harvey 2003; Venet and Baranes 2012; Wang 2013). The space era started with the Russian Sputnik launch in 1957. Although a European space programme did not get underway until the 1960s, Europeans were prominent in the development of space science and space-related technologies. Fundamental contributions to the space sector were the basic physics, mechanics and orbital theories of Italian physicists, such as Edoardo Amaldi and Enrico Fermi's group, and the German/American aerospace engineer Wernher von Braun. A number of issues helped to push Europe towards collaboration in this field at that time. One European studies theory posits a functional cooperation frame to explain EU integration.

Functionalism (Dyson and Konstadinides 2013: 115–130; Prange-Gstöhl 2010: 4, 188; Sheehan 2007: 72–90; Lane 2013: 11–33) appeals as a rational explanation in a context where the desire for peace and prosperity, linked with the Second World War, have ceased to be a driver for integration. According to this theoretical approach, during the process of integration different national interest groups shifted their loyalty away from national interests towards supranational European institutions. For instance, Germany and France were proactive in the establishment of the ESA and the initial launcher programmes. The functional explanation for this is that some national groups recognised that, compared with national action, the newly formed supranational institution was a better instrument to achieve outcomes in their interests. This led to the establishment of initial elite groups holding pan-European ideas and norms under the ESA; these groups endeavoured to persuade the national elites to turn their loyalties towards European space cooperation above the national level.

Frame analysis offers seminal insights into the understanding of EU integration in the field of space where different frames coexist, e.g. political international relations and industrial development. It enables the uncovering of the will that underlies the actions and behaviours of the EU (Mörth 2000: 175). The establishment of the two main European flagship space programmes,

Copernicus and Galileo, significantly raised the interests of all Member States to develop their space capacities and related trades, at least in downstream applications (Boin *et al.* 2009: 83–84). Frame analysis is therefore a useful tool for assessing the extent and features of European integration in space. This is despite the fact that some authors posit that longitudinal studies are necessary for an in-depth understanding, which is not yet the case for European space policy as a result of its recent entry into the integration debate (Dudley and Richardson 1999: 246). In the process of EU integration, priorities play a role at both the internal and external levels. The contest over priorities plays out through international relations between Member States inside Europe and also between the EU, encompassing the will of Member States, and the rest of the world. At the same time, each EU Member State has its own direct and unique international relations with the rest of the world. This chapter considers the international relations of the EU with third countries or in international forums. It does not address the international relations of the EU with Member States and the ESA, nor the international relations of each Member State with non-EU international actors.

International space dialogues

At present, the EU has established seven space dialogues with the USA, Russia, China, South Africa, Africa, Japan and Brazil. There are plans for Latin America to also become a dialogue partner. The Commission exercises its powers in conducting space dialogues with the collaboration of the European External Action Service. The ESA and the European Organization for the Exploitation of Meteorological Satellites are closely associated with these dialogues, in addition to EU agencies operating as consulted entities, e.g. the European Research Council, the European Defence Agency and the Executive Agency for Small and Medium-sized Enterprises (Commission 1997a, 1997b, 1998, 2013). These space dialogues mainly take two forms: first, ad hoc agreements with third countries in the context of programmes, such as Galileo, Copernicus or the Framework Programme for research and development; and second, broader, non-binding collaboration in areas such as earth observation, space exploration and the protection of space infrastructure. The rationales behind each dialogue vary because of the differing content – for example, security or technological, scientific and business perspectives. In particular, since the establishment of the two main European space flagships, Galileo and Copernicus, the EU now has its own space assets and thus can deal with third countries as a service provider of satellite navigation and earth observation applications. This helps to strengthen the industrial element of international cooperation. In addition, international cooperation in space should also support the promotion of European values through space-based projects focused on environmental protection, climate change, sustainable development and humanitarian action (ESA 2007: 142; Commission 2011: 9–10).

In conclusion, the European Commission, through the European External Action Service, is leading the space dialogues. They include the ESA as a member of the delegation, benefiting from its long tradition of scientific and technological cooperation with third countries. In addition, particular topics require coordination with other European entities, appointed on an ad hoc basis.

Space dialogue with the USA

The space dialogue with the USA is the most comprehensive and addresses all space-related concerns, including those related to security and defence, in a synergetic configuration with space issues (Steffenson 2005: 6; Delcour 2013: 84; Al-Rodhan 2012: 111; Sadeh 2013: 57; Madders 2006: 226). More importantly, different theoretical perspectives focus on the bilateral engagement of the USA with the EU, rather than presenting the EU as an additional player in bilateral relations between the USA and single European Member States. The dialogue has seen a number of concrete achievements – for example, in the areas of satellite-based navigation and the interoperability between the global positioning system and Galileo, earth observation via the partnership of the European Organisation for the Exploitation of Meteorological Satellites (EUMETSAT) with the US National Oceanic and Atmospheric Administration or, more recently, the decision allowing US public bodies to participate in EU space research projects. This achievement reflects the high level of dialogue between the EU and the USA, which has increased through the yearly EU–US Summit since 1995. The 2014 Summit reported a specific item on the space domain at number 10 of the Joint Statement, agreeing:

> to expand cooperation in research, innovation and new emerging technologies, and protection of intellectual property rights as strong drivers for increased trade and future economic growth. Our collaboration in the space domain also contributes to growth and global security, including on an International Code of Conduct for Outer Space Activities. We will combine wherever possible our efforts as we did in the Transatlantic Ocean Research Alliance and through the GPS/Galileo agreement. The Transatlantic Economic Council will continue its work to improve cooperation in emerging sectors, specifically e-mobility, e-health and new activities under the Innovation Action Partnership.
>
> (EU–US 2014)

The EU and the USA are mutually represented by delegations in Washington, from the EU side, and in Brussels, from the US side. Europe–US interaction coordinates bilateral cooperation, typified by the technological and scientific cooperation between NASA (the US national space administration) and the national European space agencies, such as Centre National d'Etudes Spatiales (the French space agency), Deutsches Zentrum für Luft- und Raumfahrt (the German space agency) and Agenzia Spaziale Italiana (the Italian space agency).

Space dialogue with Russia

Both Russia and the EU attach strategic importance to space and this is reflected in the priority given to space in the Partnership for Modernisation between the EU and Russia (Vinhas de Souza 2008: 45–48; Johnson and Robinson 2005: 14; Laursen 2009: 122; Gower and Timmins 2009: 100; Antonenko and Pinnick 2005: 69). The main topics in the EU–Russia space dialogue are the compatibility and interoperability between Galileo and GLONASS (the Russian satellite navigation system) and Russia's participation in EU space research projects, such as in the field of earth observation. The EU–Russia space dialogue also builds on the historical cooperation between the ESA and Roscosmos (the Russian space agency), covering topics ranging from launchers and scientific missions to manned spaceflight. This space dialogue was established in 2006 by the European Commission, represented by the Directorate-General for Enterprise and Industry (DG ENTR), the ESA and Roscosmos (EU–Russia 2006). Subsequently, seven working groups have been established covering all fields of space activity. Three approaches to cooperation can be identified: first, synergies and cooperation involving similar space assets such as GLONASS and Galileo, Soyuz and Ariane; second, coordination for the sharing of assets belonging to only one party, such as Copernicus of the EU and the International Space Station modules and manned launch systems held by Russia; and third, ensuring mutual policy consultations by presenting a consensus at common international forums, such as the Global Earth Observation System of Systems (GEOSS) and the International Space Exploration Coordination Group. The example of Russia shows that self-interest in controlling neighbouring issues and the scientific and technological content of the cooperation are handled by the ESA and Roscosmos in the fields of launching systems and the International Space Station.

Space dialogue with China

A key sign of cooperation between the EU and China was the EU–China summits from 1998 onwards. Earlier relationships were conducted under the economic and trade joint committee in the 1980s (Wiessala *et al.* 2009: 158; Gill and Murphy 2008: 25–27; Beneyto *et al.* 2013: 50; Schrogl *et al.* 2011: 38, 107; De Wilde *et al.* 2012: 268–271). Chinese institutions have already been involved in a major project under different EU Framework Programmes on research to measure the quality of air in Chinese cities. Although the EU–China space dialogue was only initiated in August 2012, with the participation of the ESA, China represents an important opportunity for the EU to engage in key issues related to space, including space exploration, satellite navigation, earth observation and space science. In addition, China is also an important market for European telecommunications satellites. The EU–China Summit 2012 explicitly mentioned space as a domain for cooperation, highlighting the coordination to be undertaken in the Global Navigation Satellite System field and in outlining a transparent and comprehensive plan for earth observation. The Summit recommended establishing a roadmap identifying cooperation projects and actions of mutual interest (EU–China 2012). The EU–China 2020 Strategic Agenda for Cooperation established three pillars: exchange of information in the field of earth observation, identification of a long-term plan of mutual interests and the promotion of consultations regarding Galileo and BeiDou (the Chinese satellite navigation system), Copernicus and China's earth observation programme and the development of human spaceflight technologies. In addition, commercial and industrial prospects are to be supported by each party (EU–China 2013). The most recent EU–China Summit in 2014 added additional fields of cooperation related to space weather, near-earth objects, solar system exploration and space science. In the case of China, space-related cooperation has been conducted as a coordination action between the EU and China, rather than bilateral action between national governments. China and European partners have significantly stepped up space-related cooperation in recent years, especially in the areas of earth observation satellites and training initiatives involving scientists from China visiting Europe and vice versa.

Space dialogue with South Africa

The first EU–South Africa Summit, held in 2008, identified space as an item for their future bilateral cooperation. Accordingly, the annual EU– South Africa Dialogue on Space Cooperation was initiated in January 2009. The dialogue involves the EU's DG ENTR, South Africa's Department of Science and Technology, the ESA and, since 2011, South Africa's National Space Agency (Box *et al.* 2006: 45; Boening *et al.* 2013: 113; Sicurelli 2013: 119). South Africa is particularly relevant because it hosts the Square Kilometre Array radio telescope project, one of the largest space installations in the world, which will be used to observe different parts of the sky simultaneously. The space dialogues have identified three priorities: the development of ground segments for Copernicus and one of the ground stations for Galileo; the extension of the coverage of the European Geostationary Navigation Overlay System to southern Africa; and a data exchange system for earth observation, focusing on water monitoring/management and space-based observations for managing malaria.

In addition, the EU and South Africa have been jointly cooperating under the EU Framework Programmes, the Committee on earth Observation Satellites and GEOSS. A substantial collaboration exists between the European Joint Research Centre and South African institutions on specific topics, including desertification, soil mapping, water management, crop monitoring and agriculture statistics, land management, biodiversity, and the production and validation of remote sensing parameters. Forthcoming cooperative initiatives are expected to increase the exchange of personnel such as engineers and entrepreneurs.

Space dialogue with Africa

With regard to space, the EU has very comprehensive interests in Africa as a result of challenges involving a range of specific programmes, policies and industrial competitiveness (Stokhof *et al.* 2004: 92; Van Vooren *et al.* 2013: 206; Schrogl *et al.* 2010: 230–235; Hulsroj *et al.* 2013: 100; Mangala 2013: 35; Adebajo and Whiteman 2012: 233, 267). Space has been an agenda item since the first EU–Africa Summit in the year 2000. The First Action Plan (2008–2010) and the Second Action Plan (2011–2013) of the Joint Africa–EU Strategy focused on eight priority areas of cooperation, with space constituting an element of two such areas: trade, regional integration and infrastructure; and science, information society and space. Africa is a EU cooperation and development priority. Two European space programmes have a specific African dimension: the deployment of the EGNOS navigation systems (the European space-based augmentation of the US global positioning system) in Africa and the Global Monitoring for Environment and Security and Africa programme engaging Africans in the use of earth observation services for their own needs. Other priorities under discussion with Africa include the creation of an African space agency and the establishment of space research centres at the Pan-African University. In Europe, the European Commission directly cooperates with the African Union on various levels, with the involvement of the ESA and EUMETSAT. African actors include organisations operating at a continental level (such as the African Union), regional organisations (typically the Regional Economic Communities) and technical organisations, such as African space agencies and other agencies and institutes specialising in particular areas.

Space dialogue with Japan

The EU–Japan Summit started in 1991 and, since then, several sector dialogues have taken place with Japan and the broader Asian Pacific region (Hook *et al.* 2013: 212–213; Keck 2013: 24; Mykal 2011: 151–154). Industrial cooperation was established even earlier, in 1987, with the creation of a joint venture, the EU–Japan Centre for Industrial Cooperation, between the European Commission and the Japanese Ministry of Trade and Economy. The centre has the mission of enhancing all forms of industrial, trade and investment cooperation between Japan and the EU and to strengthen the technological capabilities and competitiveness of the European and Japanese industrial systems. The space domain has not seen a strong contribution from either side and a specific space dialogue was only established at the 2014 EU–Japan Summit. The first space dialogue was held in Tokyo in October 2014 and future dialogues will be held on an annual basis. The areas for collaboration include: launch systems, with a 'coopetition' (created from competition and cooperation) approach to making the two regional technologies compatible; the field of earth observation; and forthcoming research and development activities in the areas of robotics and electronic propulsion on board satellites. Japan is willing to open a long-term collaboration with the EU because the EU's industry-driven space policy approach matches its own goals more closely than the policy approach of the USA. However, these two players do not have a long-standing relationship in joint space endeavours. Therefore research and development cooperation will be the mechanism of choice for mutually beneficial development.

Space dialogue with Brazil

The EU–Brazil Summit (Hout 2013: 126–129; El-Agraa 2007: 71; Soderbaum and Van Langenhove 2013: 20) started in 2007, with an ad hoc space policy dialogue beginning after the 2010 Summit. These annual meetings involve DG Transport (MOVE) of the European Commission and the ESA from the EU side, and, from the Brazilian side, the Ministry of Science, Technology and Innovation and the Brazilian Space Agency. Both parties signed a letter of intent in 2011, although the dialogue must still be re-activated annually. It was agreed that a steering board should supervise the dialogue and several working groups established. The areas of interest are: Earth observation and earth science, the contribution to the Group on earth Observation and the Committee on earth Observation Satellites, Global Navigation Satellite Systems), satellite communications, and space science and exploration (EU–Brazil 2012).

International cooperation in the space sector

The comprehensive landscape of international cooperation involving the EU in the space sector is mainly based on a bilateral approach – except for the collaboration with the African Union. Only the Asia Pacific region is missing from a comprehensive global scenario for European space cooperation, with interactions with Asia occurring through relations with the two competing nations of China and Japan. All the space dialogues have been conducted with the support of the ESA as a member of the EU delegation. There is an average gap of 15 years between the first summit and the first space dialogue. The shortest gap has been with China, with the EU–China Summit starting in 2000 and the related space dialogue starting in 2008. In contrast, the longest gap has been with Russia, with which country the summit was one of the earliest, commencing in 1982, although the space dialogue did not get underway until 2006.

European international relations in space: the framing approaches

To understand the magnitude of space policy as a driver for European integration in the context of international relations, the rest of this chapter investigates the work behind these space dialogues in an attempt to understand the interests and values on which they are based. Framing theory provides very useful tools of investigation with which to depict how the actors move to achieve their objectives. The analysis in this chapter is based on communications, notes and agenda items of the main institutional stakeholders, such as the European Commission, the European Parliament, stakeholders' consultations, Space Councils and similar documents. Table 12.1 shows the main items analysed for these purposes; there is no disclosure of confidential items in this chapter.

Table 12.1 Key documents from European institutions reporting the role of international relations in the space sector

Document	Title	Remarks on international relations profiles
EC, COM(1996) 617, 4 December 1996 – EP Resolution	The European Union and Space: Fostering Applications, Markets and Industrial Competitiveness	• East–West barriers • Eastern Europe can see space as a facilitator for integration • Rest of the world: USA, Russia, China, Japan, Mediterranean area
EC, COM(2000) 597, 27 September 2000 – EP Resolution	Europe and Space – Turning a New Chapter	• Cooperative scientific space missions • International trade and market access • Coordination of a European position versus United Nations • Need for ad hoc international relations on a bilateral basis for Galileo and GMES
EP Resolution A5–0119/2000	Coherent European Approach for Space	• Collaboration with USA, Russia, China and Japan for developing launch vehicles and a GPS-like system not compromised by national security considerations
EC, COM(2001) 718, 7 December 2001	Towards a European Space Policy	• Need for EU unison • Partnership with Russia through the ESA • EU with Canada and USA through the ESA for the International Space Station • Enlargement of the ESA and the EU to Eastern countries
EC, COM(2003) 17, 21 January 2003	Green Paper: European Space Policy	• Opting for international relations with leading space powers (USA, Russia) for major instruments and encompassing missions, including manned flight • Considering new space powers as Japan, China, India, Brazil and Ukraine
EC, COM(2003) 673, 11 November 2003	White Paper: Space as a New European Frontier for Expanding Union – an Action Plan for Implementing the European Space Policy	• Develop international partnerships: the benefits of establishing a strategic partnership with Russia, of maintaining and developing Europe's long-standing partnership with the USA and of exploiting other emerging cooperation with new space powers such as Brazil, China, India, Japan and Ukraine
Third Space Council, 17 November 2005	–	• Presidency discussion on first reflections on international relations

Fourth Space Council, 25 May 2007	—	• International relations under United Nations • International relations for attracting partners in Galileo and Copernicus (formerly GMES) • International relations for including other fields, such as sustainable development, climate change, humanitarian aid, foreign and security • Partnership on space with Africa
Seventh Space Council, 26 November 2010	—	• International relations are vital for space • International relations as promoter of European values through space for environment, climate change, sustainable development and humanitarian action • International relations of EU with Russia and China • International relations for EGNOS and Copernicus providing support to Africa
EC, COM(2011) 152, 4 April 2011	Towards a Space Strategy for the European Union that Benefits its Citizens	• International relations of EU under UN: code of conduct • International relations as multilateral form for GEOSS

The first finding concerns the interlocutors of the EU in the space domain. There are international multilateral actors, such as the United Nations for the Code of Conduct to enhance the safety, security and sustainability of activities in outer space and Galileo coordination, and the GEOSS for Copernicus issues. EU framing dynamics have shaped international relations where the main space flagship programmes have been approved and already deployed. In the 1990s, the EU was looking for partners to work with on Galileo and Copernicus and coordinated with countries that had similar space-based systems, while promoting downstream development with countries without their own space assets.

The beginning of EU space affairs saw the intention to propose space as content for cooperation with Eastern European countries on their integration paths. Cooperation is intended to support the integration of these countries into trans-European networks – for example, information and communication technology infrastructures, smart grids or equipment standards (see Chapter 9). This also means that, for the continent, international cooperation started within European integration (Commission 1996: 5–10). This Communication of the Commission on space matters focuses attention on the need for a coordinated and concentrated European space strategy for international cooperation as a result of the change from a bipolar geopolitical context to a more globalised and competitive world environment. Attention is also given to other issues – for example, industrial and technological cooperation with the established space nations (the USA, Russia, China and Japan) and collaboration with emerging countries such as India and Brazil. The Euro-Mediterranean area is a region of space cooperation where European industry, supported by the EU, can serve the information and communication needs of these nations. The Communication of 1999 (Commission 1999: 17–21) places the development of an independent satellite navigation system inside the scope of international cooperation, starting with cooperation with existing US and Russian systems, through to the coordination of the augmentation systems, such as EGNOS (the European augmentation system for navigation) with the Japanese Multi-functional Satellite Augmentation System. These messages are redundantly supported in different policy statements of European institutions (Commission 1997a, 1997b; European Parliament 2000a, 2000b; Council 2002) and reported in the Resolution of the European Parliament:

> urges the further development of international collaboration in space activities, in particular with the Russian Federation, the USA, China and Japan, but also with less developed countries for which the European Union could provide affordable access to space, and asks the Commission to arrange a conference to explore the possibilities of such co-operative ventures with representatives of the above four space powers, covering scientific, technological, industrial or economic aspects, such as the new international orbiting space station.
>
> (European Parliament 2002)

The Green Paper of the European Commission on space policy outlines the fundamental elements of international cooperation to converge on a single European vision of space 'speaking with one voice' (Commission 2003a: 15). International cooperation with the main space powers is envisaged to avoid the duplication of investment in research and development. In addition, international relations are supportive of clarifying matters of international trade such as regulation, standardisation and non-tariff protection.

There are also elements of intended cooperation with India and the Ukraine (Commission 2000, 2001), where a proper space dialogue has not yet been reached. The case of India sounds a paradoxically missed opportunity for concrete action on space cooperation (Wülbers, 2010: 76–83). Relations between the EU and India benefit from a long-standing relationship going back to the early 1960s. These relations then matured in the Joint Political Statement of 1993 and the Co-operation Agreement of 1994, where space is an item of cooperation in the field of earth and space-based telecommunications (EU–India 1994). In 2000, these two parties established an annual summit; the Fifth EU–India Summit of 2004 explicitly mentioned space as an area of cooperation. At this summit, the two parties expressed their will to enhance collaboration between the ESA and the India Space Agency and the EU expressed its interest in the Indian unmanned lunar exploration mission, Chandrayaan-1 (EU–India, 2004).

In the following year, the parties enlarged the areas of cooperation to 'communications, meteorology, navigation, life and material sciences under microgravity conditions, space exploration, space sciences and any other area relevant fields to the respective space programmes' (EU–India, 2005). In addition, space is an element of the EU–India Economic Policy Dialogue and Co-operation, established in the EU–India Country Strategy Paper for India 2007–2013, although no specific action is defined in the space field. None of these elements has yet led to any specific space dialogue between the EU and India.

The core discussion about the role of international relations was shaped in the third Space Council in 2005. Here, international cooperation was seen as a driver for the European space strategy, with benefits for Europe and the European market being the key criteria for ranking the prioritisation of international partnerships (Space Council 2005). The Fourth Space Council in 2007 enhanced the role of international relations with international organisations – that is, the United Nations and particularly the Committee on the Peaceful Uses of Outer Space – in establishing and maintaining the rule of law in using and exploring outer space (Space Council 2007).

In 2008, the Commission mentions the role of international relations with regard to improving transparency and the need for the coordination of a single European voice from all stakeholders dealing with the space dialogues (Commission 2008). The priority is put on coordinated actions boosting the market competitiveness of European industry by using space systems for sustainable development – for example, in Africa – and contributing to the practical implementation of European space programmes. The European partnership on space with Africa was officially established under the Seventh Space Council in 2010, with the provision of space-based applications (EGNOS, Copernicus and satellite communications) helping to enable Africa to achieve the Millennium Development Goals (Space Council 2010). The communication from the Commission in 2011 shapes a clear comprehensive statement on the role of international relations as a substantial tool for assuring the sustainability of European investments in space assets such as Galileo and Copernicus. The space dialogues of the EU with the ESA will enhance cooperation for the mutual benefit of the partners to integrate the external actions with the internal priorities of the EU, such as socio-economic growth and industrial competiveness. Thus the contents of international collaboration are sharing policies for the utilisation of assets, data, signals and frequencies with the space powers (the USA, Russia, China and Japan), while supporting service provision for the needs of developing areas such as Africa and Brazil.

The findings of this chapter show that there are two main framing approaches within the overall environment of European space policy where space and international relations mutually influence each other. The space environment changed in international politics in the early 1990s when political and economic globalisation had an impact on the space domain. There has been an increasing number of actors engaging in space activities and a diversification of their nature from public to private actors, non-governmental organisations and international organisations. More generally, space activities are mirroring the major issues and debates at stake in international relations: competition versus cooperation; unilateralism versus multilateralism; polarity of the international system versus globalisation; international trade; and global governance. The enlargement of these new types of players – industrial companies and emerging space nations – affected the functioning of international relations. Thus when the geopolitical context is tense, it is valuable to use space as a powerful element in the discussion (Commission 2003b). An example of this is how the EU built security, transparency and confidence through two complex assets, – Galileo and Copernicus – in dealing with the main space-faring nations with their own space-based systems, namely the USA, Russia, China and Japan.

What is missing is a comprehensive European space policy with a specific plan for international coordination with the USA to balance the emerging competition from Russia, Ukraine and China in the launch sector. Faced with these developing countries, the international coordination problems of the EU are signs of a lack of a common European voice and this is a serious handicap to achieving a strong market position. When the market needs in specific regional areas are not fully satisfied as a result of the lack of technological expertise or industrial know-how, and when the geopolitical stake is not binding, then the industrial lobbies usually take the lead. Naturally they foster their own frames. In these cases, the EU behaves as a promoter of its own industries, mainly in the field of downstream applications, towards emerging space nations such as Africa, South Africa and Brazil.

Conclusions

There is no doubt that the European space policy is industry-oriented; space is a catalyst of industrial competitiveness. It feeds the ambition for European technological leadership. When negotiating over space affairs with established space powers, the industrial content is partially overtaken by geopolitical priorities. The wording of European policy statements often refers to the aim to strengthen the industrial sector through the promotion of socio-economic growth, sustainable development and the sustainability of space activities.

In a broader context, the EU is shaping its international relations and international cooperation in space by focusing on both industrial interests and geopolitical values. International space cooperation serves European interests regarding its space policy and its more general foreign policy objectives. On the one hand, space cooperation is a pragmatic tool to implement ambitious programmes such as space exploration and the International Space Station; on the other hand, international space activities foster a climate of political cooperation and help in responding to Europe's political priorities such as climate change, sustainable development, humanitarian aid and security in the broad sense. At the same time, Europe's space activities illustrate its role and position in the international system. Europe is still an international actor under construction and is in search of its specific international identity as neither a state nor a classical intergovernmental organisation, but an actor of its own kind.

Space has been a catalyst for EU internal cohesion – for example, with New Member States from Eastern Europe in the 2000s. At the external level, messages coming from the EU are still vague as they emerge as generic declarations of intent. On the international stage, the EU needs to find an adequate balance between cooperation and autonomy, making itself a credible cooperation partner for the established space-faring nations. The ambiguity of EU identity remains even in these negotiations, in which the EU puts across the sum of the interests of its Member States. We are still far from a unique European space policy expressed in a single voice on the international stage. European international relations in space remain in a fragile balance between cooperation and competition because the different approaches adopted by the EU in framing its international relations express the aim of making the EU an actor in the international landscape in its own right, but are also the expression and influence of its internal characteristics.

Another feature of the international relations of the EU in the space sector is that the shape of international cooperation has been mainly inherited from the ESA, where international relations activities are more closely tied to its programmes and less to broader policy considerations. Because of this, there is no clear road map for the international relations of the EU regarding space, with most relationships established on an opportunistic, pragmatic basis to support existing programmes. Past evidence shows that space cooperation almost always starts with science, where costs are high and collaboration is known to reduce costs. The purpose is to enhance the scientific value of missions, share risks, combine competencies and promote open data policies for scientific purposes.

When the EU has its own space facilities and when it can express its normative political values, then its international relations will be shaped by a strategy that is then implemented through appropriate programmes. The history of EU international relations in the field of space thus presents itself as a proliferation of programmes in need of a strategy.

References

Adebajo, A. and Whiteman, K. (2012) *The EU and Africa: From Eurafrique to Afro-Europa*. London: Hurst.

Al-Rodhan, N.R.F. (2012) *Meta-geopolitics of Outer Space: An Analysis of Space Power, Security and Governance*. Basingstoke: Palgrave Macmillan.

Antonenko, O. and Pinnick, K. (2005) *Russia and the European Union: Prospects for a New Relationship*. London: Routledge.

Beneyto, J.M., Song, X. and Ding, C. (2013) *China and the European Union: Future Directions*. San Pablo: Fundación Universidad de San Pablo.

Boening, A., Kremer, J.-F. and van Loon, A. (2013) *Global Power Europe – Vol. 2: Policies, Actions and Influence of the EU's External Relations*. New York, NY: Springer Science and Business Media.

Boin, A., 't Hart, P. and McConnell, A. (2009) 'Crisis exploitation: political and policy impacts of framing contests'. *Journal of European Public Policy*, 16(1): 81–106.

Box, L. and Engelhard, R. (2006) *Science and Technology Policy for Development: Dialogues at the Interface*. London: Anthem.

Commission (1996) *The European Union and Space: Fostering Applications, Markets and Industrial Competitiveness. Communication from the Commission to the Council and the European Parliament.* COM(1996) 617, 4 December 1996.

Commission (1997a) *European Union – Russia Space Dialogue: Co-operation and Coordination in Space Related Activities.* Press Release IP/97/523, 15 June 1995.

Commission (1997b) *The European Union and Space: Fostering Applications, Markets and Industrial Competitiveness.* Report A4–0384/97, 3 December 1997.

Commission (1998) *New Impetus for the 'EU–Russia Space Dialogue'.* Press Release IP/98/709, 28 July 1998.

Commission (1999) *Towards a Coherent European Approach for Space.* SEC (1999) 789, 6 June 1999.

Commission (2000) *Europe and Space Turning to a New Chapter. Communication from the Commission to the Council and the European Parliament.* COM(2000) 597, 27 September 2000.

Commission (2001) *Towards a European Space Policy.* COM(2001) 718, 7 December 2001.

Commission (2003a) *Green Paper: European Space Policy.* COM(2003) 17, 21 January 2003.

Commission (2003b) *White Paper: Space as a New European Frontier for Expanding Union – An Action Plan for Implementing the European Space Policy.*, COM(2003) 673, 11 November 2003.

Commission (2008) *European Space Policy Progress Report.* COM(2008) 561 final, 11 September 2008.

Commission (2011) *Towards a Space Strategy for the European Union that Benefits its Citizens.* COM(2011) 152, 4 January 2011.

Commission (2013) *Key Facts on the Joint Africa–EU Strategy.* Memo 23, April 2013.

Council (2000) *European Space Strategy.* 2000/C 371/02, 23 December 2000.

De Wilde, T., Defraigne, P. and Defraigne, J.C. (2012) *China, the European Union and the Restructuring of Global Governance.* Cheltenham: Edward Elgar.

Delcour, L. (2013) *Shaping the Post-Soviet Space? EU Policies and Approaches to Region-Building.* Farnham: Ashgate.

Dudley, G. and Richardson, J. (1999) 'Competing advocacy coalitions and the process of "frame reflection": a longitudinal analysis of EU steel policy'. *Journal of European Public Policy,* 6(2): 225–248.

Dyson, T. and Konstadinides, T. (2013) *European Defence Cooperation in EU Law and IR Theory.* Basingstoke: Palgrave Macmillan.

El-Agraa, A. (2007) *The European Union: Economics and Policies.* Cambridge: Cambridge University Press.

European Parliament (2000a) *Coherent European Approach for Space.* Resolution A5–0119/2000.

European Parliament (2000b) *Towards a Coherent European Approach for Space.* Resolution A5–0119/2000, Minute 18 May 2000.

European Parliament (2002) *Europe and Space: Turning to a New Chapter.* Resolution P5–TA (2002) 0015.

European Space Agency (2007) *Handbook.* Washington DC: International Business

European Union–Brazil (2012) *14th Meeting of the Brazil–European Union Joint Committee.* Joint Communication, Brasilia, 1 June 2012. Available from: http://eeas. europa.eu/brazil/docs/20120608_14th_meeting_br_eu_jc_en.pdf. Last accessed 22 May 2015.

European Union–China (2012) *15th EU–China Summit, Towards a Stronger EU–China Comprehensive Strategic Partnership,* 20 September 2012. Available from: http:// europa.eu/rapid/press-release_MEMO-12-693_en.htm?locale=en. Last accessed 22 May 2015.

European Union–China (2013) *China–EU 2020 Strategic Agenda for Cooperation Released at 16th China–EU Summit.* Press Release, 131123/01, 23 November 2013. Available from: http://eeas.europa.eu/statements/docs/2013/131123_01_en.pdf. Last accessed 22 May 2015.

European Union–India (1994) *Cooperation Agreement between the European Community and the Republic of India on Partnership and Development – Declaration of the Community Concerning Tariff Adjustments – Declarations of the Community and India.* L. 223/24, 27 August 1994. Available from: http://ec.europa.eu/world/agreements/prepareCreateTreatiesWorkspace/treatiesGeneralData.do?step=0&redirect=true&treatyId=352. Last accessed 22 May 2015.

European Union–India (2004) *Fifth India–EU Summit – Joint Press Statement.* 14431/04, Presse 315, 8 November 2004. Available from: http://europa.eu/rapid/press-release_PRES-04-315_en.htm?locale=en. Last accessed 22 May 2015.

European Union–India (2005) *The India–EU Strategic Partnership: Joint Action Plan.* 11984/05, Presse 223, 7 September 2005.

European Union–Russia (2006) *Russia–EU Dialogue on Space Cooperation: Joint Statement, Brussels, 10 March 2006.* Available from: www.ec.europa.eu/enterprise/policies/ international/files/eu_russia_space_dialogue_en.pdf. Last accessed 22 May 2015.

European Union–US (2014) *Joint Statement EU–US Summit.* 140326/02, 26 March 2014. Available from: www.eeas.europa.eu/statements/docs/2014/140326_02_en.pdf. Last accessed 22 May 2015.

Gill, B. and Murphy, M. (2008) *China–Europe Relations: Implications and Policy Responses for the United States: A Report of the CSIS Freeman Chair in China Studies.* Washington, DC: Center for Strategic and International Studies.

Gower, J. and Timmins, G. (2009) *Russia and Europe in the Twenty-First Century: an Uneasy Partnership.* London/New York, NY: Anthem Press.

Harvey, B. (2003) *Europe's Space Programme: To Ariane and Beyond.* Dublin: Springer Science and Business Media.

Hook, G.D., Gilson, J., Hughes, C.W. and Dobson, H. (2013) *Japan's International Relations: Politics, Economics and Security.* London: Routledge.

Hout, W. (2013) *EU Strategies on Governance Reform – Between Development and State-building.* London: Routledge.

Hulsroj, P., Pagkratis, S. and Baranes, B. (2013) *Yearbook on Space Policy 2010/2011: The Forward Look.* Vienna/New York: Springer Science and Business Media.

Johnson, D. and Robinson, P.F. (2005) *Perspectives on EU–Russia Relations.* London: Routledge.

Keck, J., Vanoverbeke, D. and Waldenberger, F. (2013) *EU–Japan Relations, 1970–2012: From Confrontation to Global Partnership.* London: Routledge.

Lane, D. (2013) *Elites and Identities in Post-Soviet Space.* London: Routledge.

Laursen, F. (2009) *The EU in the Global Political Economy.* Brussels: Peter Lang.

Madders, K. (2006) *A New Force at a New Frontier: Europe's Development in the Space Field in the Light of its Main Actors, Policies, Law and Activities from its Beginnings up to the Present.* Cambridge: Cambridge University Press.

Mangala, J. (2013) *Africa and the European Union: a Strategic Partnership.* Basingstoke: Palgrave Macmillan.

Mörth, U. (2000) 'Competing frames in the European Commission – the case of the defence industry and equipment issue'. *JEPP*, 7(2): 173–189.

Mykal, O. (2011) *The EU–Japan Security Dialogue: Invisible but Comprehensive.* Amsterdam: Amsterdam University Press.

Prange-Gstöhl, H. (2010) *International Science and Technology Cooperation in a Globalized World: the External Dimension of the European Research.* Cheltenham: Edward Elgar.

Sadeh, E. (2013) *Space Strategy in the 21st Century: Theory and Policy.* London: Routledge.

Schrogl, K.-U., Mathieu, C. and Peter, N. (2010) *Yearbook on Space Policy 2007/2008: From Policies to Programmes.* Vienna/New York, NY: Springer Science and Business Media.

Schrogl, K.-U., Baranes, B., Venet, C. and Rathgeber, W. (2011) *Yearbook on Space Policy 2008/2009: Setting New Trends.* Vienna/New York, NY: Springer Science and Business Media.

Sheehan, M. (2007) *The International Politics of Space.* London: Routledge.

Sicurelli, D. (2013) *The European Union's Africa Policies: Norms, Interests and Impact.* Farnham: Ashgate.

Soderbaum, F. and Van Langenhove, L. (2013) *The EU as a Global Partner: The Politics of Interregionalism.* London: Routledge.

Space Council (2005) *Third Space Council.* 14499/1/05, 17 November 2005. Available from: http://register.consilium.europa.eu/doc/srv?l=EN&f=ST%2014499%202005%20REV%201. Last accessed 22 May 2015.

Space Council (2007) *Fourth Space Council.* 10037/07, 25 May 2007. Available from: http://ec.europa.eu/enterprise/newsroom/cf/_getdocument.cfm?doc_id=1972.   Last accessed 22 May 2015.

Space Council (2010) *Seventh Space Council.* 16864/10, 26 November 2010. Available from: http://register.consilium.europa.eu/doc/srv?l=EN&f=ST%2016864%202010%20 INIT. Last accessed 22 May 2015.

Steffenson, R. (2005) *Managing EU–US Relations: Actors, Institutions and the New Transatlantic Agenda.* Manchester: Manchester University Press.

Stokhof, W., van der Velde, P. and Hwee, Y.L. (2004) *The Eurasian Space: Far More Than Two Continents.* Singapore: Institute of Southeast Asian Studies.

Van Vooren, B., Blockmans, S. and Wouters, J. (2013) *The EU's Role in Global Governance: The Legal Dimension.* Oxford. Oxford University Press.

Venet, C. and Baranes, B. (2012) *European Identity through Space: Space Activities and Programmes as a Tool to Reinvigorate the European Identity.* Vienna/New York, NY: Springer Science and Business Media.

Vinhas de Souza, L. (2008) *A Different Country: Russia's Economic Resurgence.* Brussels: CEPS.

Wang, S.C. (2013) *Transatlantic Space Politics: Competition and Cooperation above the Clouds.* London: Routledge.

Wiessala, G., Wilson, J.F. and Taneja, P. (2009) *The European Union and China: Interests and Dilemmas.* Amsterdam: Rodopi.

Wülbers, S.A. (2010) *The Paradox of EU–India Relations: Missed Opportunities in Politics, Economics, Development Cooperation, and Culture.* Lanham: Lexington Books

Part IV

Policy process

Implementing space policy

13 Lessons from Galileo for future European public–private partnerships in the space sector

Jakob Feyerer

Introduction

In March 2002, the European Council gave the green light to the Transport Council to fund and develop a European Global Navigation Satellite System (GNSS), named Galileo. This became the first EU level public–private partnership (PPP) in cooperation with the European Space Agency (ESA) (Barcelona European Council 2002: 16). Galileo was of high strategic and economic interest to the EU. This project would provide the EU with independence from the US global positioning system (GPS) in safety-critical transport areas and public transport infrastructure. It would avoid possible charges for the EU's use of the US GPS and protect the EU from potential scenarios of restricted access in case of conflict, as occurred in the Kosovo war in 1998–1999. Moreover, Galileo was supposed to increase the EU's share in the growing world market for GNSS applications and services. The European Commission recommended developing Galileo as a PPP as it saw the general advantages of this concept in the provision of private funding, value for money[1] and increased efficiency through the use of private expertise (Commission 1999: 4–5).

In June 2007, the European Transport Council came to the conclusion that the PPP had failed as a result of the inability of the companies involved to fulfil the requirements of the Galileo project. By 2007, Galileo had been delayed by five years and its budget was double that initially calculated (from €1.1 to 2.1 billion; Riccardi 2007: 2). The PPP was terminated and Galileo became a system fully controlled and funded by the EU. The analysis of the negotiations of the Galileo PPP shows that two issues were of particular importance in the negotiations: the terms of the contract and the interests of the involved actors in reaching an agreement.

A number of policy frames appeared in the negotiations for the Galileo PPP. According to Baumgartner and Mahoney (2008):

> Research based on framing theory focuses on two interrelated questions. First: how individual actors try to frame an issue, and whether or not they manage to adapt their argumentation to the recipient. The second question is what frames exist in a debate and how these frames change over time.
>
> (Baumgartner and Mahoney 2008: 444)

It can be assumed that the ability of individual actors to reframe the debate is limited. Baumgartner and Mahoney (2008: 445) point out that although, in certain situations, windows of opportunity open for individual actors, issues are usually led 'not only by institutional procedures, but also by collective policy knowledge and expertise' and are usually 'collectively defined in a stable manner' (Baumgartner and Mahoney 2008: 445). This is particularly true for a highly complex project such as Galileo. Although the two perspectives of framing theory are interrelated, the focus is on the collective frames of the Galileo process. The analysis of the negotiations for the Galileo PPP shows that decision-making was determined by a policy frame concerned with the economic advantages of the PPP concept. The dominant frame imposed other sub-frames dealing specifically with Galileo and with the design of the project as a PPP.

Based on a variety of sources, this chapter aims to trace the Galileo decision-making process and identify the central reasons for the termination of the contract negotiations. The analysis is based on: Council Resolutions and Regulations; Communications of the European Commission; reports of other European institutions such as the European Court of Auditors; and semi-structured interviews with experts – for example, with a member of the GNSS panel of the Council of the EU and with a High Representative of the European GNSS Supervisory Authority. The academic literature provides important information for analysing PPPs in the EU context. The third section describes the Galileo programme. There is a focus on the negotiations for the Galileo PPP. The concluding section shows which frames appeared in the negotiations and summarises the lessons to be learnt for future EU PPPs in the space sector.

Public–private partnerships in the EU

Bovaird (2004: 201) identifies two driving factors in the PPP concept. One factor is national fiscal problems, forcing the state to use private funding to provide public services. The second crucial factor is the growing importance of e-Government,[2] supporting closer public–private cooperation, especially in the information and communication technology sector, where the public sector needs the capital and particularly the technological know-how of private companies (Bovaird 2004: 201). The EU supports infrastructure investment to stabilise the financial sector and to cushion the impacts of the financial crisis.[3] The average size of infrastructure PPPs in the EU decreased from €217 million in 2007 to €91 million in 2009 (Kappeler and Nemoz 2010: 20–22). Nevertheless, the PPP concept remained a popular instrument to provide public infrastructure at the EU level and in some EU Member States. As pointed out by the European Commission in a trend-setting Communication, 'public–private partnerships can provide effective ways to deliver infrastructure projects, to provide public services and to innovate more widely in the context of these recovery efforts' (Commission 2009: 2). The Commission emphasised the value of the PPP concept in the development of public services and infrastructure, which would allow the strengths of the public and the private sectors to be combined (Commission 2009: 2). To support this position, the Commission claims there are various benefits to the PPP concept.[4] It argues that the PPP concept improves the performance of a project with respect to on-time and on-budget delivery compared with traditional procurement. A PPP could furthermore improve the value for money of the infrastructure project, reducing public investment costs by distributing the costs over a number of years. A PPP could improve the risk management of a project and stimulate sustainability and innovation. It could improve the position of the private sector in the long-term planning and implementation of large-scale projects and it could increase the market shares of EU companies outside the EU (Commission 2009: 3–4).

The term PPP has many different meanings and implications. A general definition to embrace all the characteristics of this term would be too broad to have any value. Any PPP definition therefore has to be tailored to a certain purpose. Many PPPs are, for example, in the areas of infrastructure, development or urban regeneration (Weihe 2008: 431–435). Although the focus of this chapter is on infrastructure PPPs, even in this category there is a broad variety of definitions. Bertrán and Vidal (2005: 390), for example, broadly describe PPPs as 'partnerships between the public sector and the private sector (industry), for the purpose of delivering a project or a service traditionally provided by the public sector'. Grimsey and Lewis (2004: xiv) define a PPP as 'a risk-sharing relationship based on a shared aspiration between the public sector and one or more partners from the private and/or voluntary sectors to deliver a publicly agreed outcome and/or public service'. De Palma et al. (2009: 3) define PPPs as 'contractual arrangements covering a long-time period (typically more than 20 years) by which public authorities assign to private operators the fulfilment of a mission of public interest'. From the perspective of the European Commission, the term PPP 'refers to forms of cooperation between public authorities and the world of business which aim to ensure the funding, construction, renovation, management or maintenance of an infrastructure or the provision of a service' (Commission 2004: 3). According to Renda and Schrefler (2006: 1), the central aim of a public–private contractual agreement is to generate more value for money than a traditional public procurement. The term indicates substantial benefits for customers and taxpayers (return on investment) (Bertrán and Vidal 2005: 391).

The Galileo PPP was negotiated as a 'design build finance operate' (DBFO) concession contract, one of the most complex forms of infrastructure PPPs. In a DBFO agreement, a service or asset is provided by the private sector according to public provisions. The private sector is responsible for, and bears the financial risk of, the design, construction and operation phase (Renda and Schrefler 2006: 5). The private partner in a DBFO is usually a consortium of firms, installing a special purpose vehicle to manage the project and the risk allocation among the involved actors (Iossa et al. 2007: 26–27). When the contract ends, the service or asset can be returned to the public sector or the contract can be renegotiated. A DBFO is very flexible and can be adapted to the individual service or asset. In a DBFO concession, the most common DBFO form:

The private investor designs, finances, constructs and operates a revenue-generating infrastructure in exchange for the right to collect the revenues for a specific period of time, generally for 25–30 years. Ownership of the asset remains with the public sector.

(Renda and Schrefler 2006: 5)

Galileo has so far been the only attempt by the EU to construct a transnational[5] EU level infrastructure PPP. To understand why the Galileo PPP negotiations failed, we need to identify those factors determining the success of a transnational EU PPP in the space sector. A study for the ESA identified the following factors:

Efficiency of risk allocation; the ability to increase project efficiency; adequate financing; the existence of clear outcome criteria; the potential of cost savings (value for money); the ability to foster innovation; contract model flexibility; maturity of the PPP model; public control of assets; contract implementation; [and a] regulatory environment.

(Bochinger *et al.* 2009: 13–14)

The factors presented by the ESA study are formulated in a very general manner, which is legitimate given the fact that the complexity of the term PPP prevents a narrow definition. These factors appear to be an attempt to define the specifics of EU PPPs in the space sector, focusing predominantly on the details of the contract. The ESA study ignores the individual interests of the involved actors, identified by the PPP literature as an important dimension with great influence on the success of negotiations for a PPP.

Hodge and Greve (2009: 37), for example, point out that the role of a governmental actor in a PPP is ambiguous and difficult, as it:

now finds itself in the middle of multiple conflicts of interest, acting in the roles of policy advocate, economic developer, steward for public funds, elected representative for decision-making, regulator over the contract life, commercial signatory to the contract and planner.

(Hodge and Greve 2009: 37)

With regard to transnational PPPs, Schäferhoff *et al.* (2009: 456) argue that PPPs are the result of the overlapping interests of actors and not of their intention to solve global challenges. The emphasis of this approach lies in the interest of the actors to gain resources they could not provide themselves. Grimsey and Lewis (2004: 160) emphasise that interaction is an essential element of the tendering process. They argue that 'the negotiations for contract fulfilment need to be managed with cooperation and forbearance' to face '[c]ompetitive tensions' (Grimsey and Lewis 2004: 160). Flinders (2005: 234) argues that PPPs change the focus of the public actors in favour of economic and against public interests. He states that these value systems are eventually incompatible. In general, it can be concluded that the literature identifies two dimensions of factors with particular importance for PPP negotiations in the space sector: the contractual agreements and the interests of the individual actors.

The Galileo project

The European Commission called for the development of Galileo[6] for three reasons. First, to avoid possible charges or fees to the European public for the use of the signal from an external system. Second, to make sure that the EU could participate in the lucrative market for satellite navigation, which 'could be worth 50 billion dollars within 7 years' (Commission 1998: iv). Third, to protect European security interests because, at the time, the only available signals were from the external military systems, the US GPS and the Russian GLONASS (Commission 1998: iv).

Galileo was intended to provide four basic services: an Open Service free of charge to users; a Safety of Life Service for critical transport vessels such as ships and airplanes; a Commercial Service with higher accuracy than the Open Service through two additional signals; and a Public Regulated Service (PRS) reserved for institutions preserving governmental security interests. The EU also participates in international cooperation for humanitarian search and rescue with the Search and Rescue Service (ESA 2010). The Council of Transport Ministers in its meeting of 9–10 December 2004 defined Galileo as a 'civil programme under civil control', emphasising the notion of a civilian Galileo, without excluding military use of the PRS (Council 2004a: 24) and leaving room for interpretation in both directions. France and other countries claimed that the PRS was an issue of national sovereignty and thus should be accessible for all EU Member States. The UK and other Member States argued that military use of the PRS falls under the Common Security and Defence Policy and should therefore be decided unanimously (Nardon and Venet 2010: 3). A member of the GNSS panel of the Council stated that, 'never was there a decision to exclude military use. You can have a military use of Galileo. It is not at all excluded. And it is one of the uses that were contemplated by the Member States.' (personal interview, 28 June 2012). In summary, the main purpose of Galileo was the civilian use and commercial exploitation of the signal. The military use of Galileo was discussed, but the negotiations for the development of Galileo as a PPP were dominated by non-military issues.

In 2011, the European Commission estimated the annual worldwide turnover of markets for satellite navigation infrastructure to be around €130 billion in 2010 and expected rapid growth of these markets, reaching an estimated turnover of around €240 billion by 2020 (Commission 2011b: 12). The Commission expected Galileo to 'generate economic and social benefits worth around €60–90 billion over the next 20 years' (Commission 2011b: 12). However, the EU Commission also noted that these potential benefits could only be generated if Galileo was delivered on time. According to the Commission, 'each year's delay will decrease the value of the benefits by 10–15 per cent owing to both the loss of revenue generated and the development of alternative solutions and competing systems' (Commission 2011b: 12).

Galileo continues to be developed as a central part of the Trans-European Networks (TEN) programme. The EU Commission describes Galileo as 'a key part of the Transport TENs and the Common Transport Policy' and as 'an essentially trans-European project, bringing direct benefits to all Member States (helping them meet their public service and international obligations with respect to providing navigation aids)' (Commission 1999: 14). The Trans-European Transport Network (TEN-T) programme was initiated in the late 1980s to support the Single European Market, together with the Cohesion and Structural Funds (PricewaterhouseCoopers 2005: 72), by constructing 'an integrated network of basic transport infrastructure, transforming the networks built under national considerations into an efficient and sustainable Europe-wide infrastructure system'. According to the EU Commission (2011a: 1), a core TEN-T network was planned 'to be established by 2030 to act as the backbone for transportation within the Single Market ... supported by a comprehensive network of routes, feeding into the core network at regional and national level ...' by 2050. Galileo is a central project of the TEN-T programme. According to the EU Commission, Galileo is:

> [a] key priority... to develop a satellite radio navigation system for civil use. Galileo will help to improve efficiency and safety in all transport modes, while at the same time guaranteeing the European Union's technological independence in this area.
>
> (Commission 2003: 6)

The European Investment Bank estimates the overall costs of TEN-T projects at €900 billion by 2020 (European Investment Bank 2009: 6). The European Commission supports the PPP concept to meet the requirements for additional funding of TEN-T projects and has therefore called for improvements in the legal framework 'particularly as regards concession rights and charging for infrastructure use'[7] and for new risk-covering mechanisms. (Commission 2003: 8).

The development of Galileo was strongly influenced by the prevailing paradigm in the EU that the technical expertise and financial resources of the private sector could significantly improve the performance of public infrastructure provision. In addition, PPPs are considered as a useful instrument to alleviate the effects of the financial crisis. In accordance with this paradigm, the EU Commission argued in favour of developing Galileo as a PPP.

The negotiations for the Galileo PPP

According to the EU Commission, 'a PPP structure will help keep costs under control since much of the risk of construction cost over-run would normally fall on the private sector. It would also reflect the fact that Galileo combines public service and commercial aspects' (Commission 1999: 18). The Council approved the PPP approach for Galileo and charged the Commission with starting the definition phase in 1999 (Council 1999: 1–3). The Commission (2000: 28) formulated a number of prerequisites for further negotiations – for example, a sound commercial plan, additional technical risk studies, or a clear public management structure. A political decision of the European Transport Council of December 2000 in Nice was considered as essential for further negotiations: 'All of these conditions should be met by 2003 provided that a firm decision on continuing the programme is taken in December 2000' (Commission 2000: 28).

The Commission presented the results of the technical definition phase in 2000, including a funding scheme for the development and validation phase (envisaged from 2001 to 2005) and an outlook to the deployment (2006–2007) and operational (from 2008) phases in a Communication to be approved by the Transport Council. This Communication emphasised the PPP design of Galileo as 'an essential factor for the success of the Galileo programme' (Commission 2000: 18–27). The Commission favoured a PPP model because 'early development of user segments was seen as the key to subsequent use of the Galileo system, if direct revenue was to be generated' (Court of Auditors 2009: 13). In 2001, the Council requested the Commission to undertake studies to develop a suitable business plan for Galileo (Council 2001a: 1–3). The conclusion of these studies was that a DBFO-type concession PPP was the most suitable model for Galileo. The PPP model for Galileo differed from a regular DBFO because the private sector was meant to build, finance and operate a system designed by the public sector (Court of Auditors 2009: 13–25).

The Commission's reasoning for a Galileo PPP was mainly based on a study developed by the accountancy firm PricewaterhouseCoopers in 2001, which strongly supported the concession model. This study was very optimistic about the overall cost–benefit ratio of Galileo and called for the development of Galileo as a PPP (PricewaterhouseCoopers 2001: 5–9). Transport Commissioner De Palacio presented the expected costs and benefits of Galileo according to the PricewaterhouseCoopers study in the Transport Council of December 2001 to convince the Council Members to decide to go ahead with the Galileo development phase. Five EU Member States[8] opposed the position of the Commission, insisting on a clear declaration of intention by the private sector to bear the vast majority of the costs for the Galileo project. They argued that the results of this study would not attract enough private capital to cover the costs of the development and deployment phases (Laeken European Council 2001). The private sector responded to this request by signing a specific memorandum of understanding for a binding commitment to participate in the Galileo project and in the funding of the development phase (Council 2001b). With this memorandum, the private sector declared its willingness to engage in the project, but this document was not a binding financial commitment.

The Transport Council rejected a clear commitment to Galileo in 2000. This delay took the ESA, the Commission and the space industry by surprise. However, the EU Member States could not agree on three central issues: the conditions of funding; security issues; and the legal framework of the programme.

Other controversial issues were the integration of the European Geostationary Navigation Overlay Service (EGNOS) into Galileo, legal liabilities in case of a malfunction of Galileo, the allocation of frequencies and the design of the terrestrial infrastructure (Lembke 2001: 9–10). The PricewaterhouseCoopers study did not result in any rapid decision-making process on Galileo, but it substantially strengthened trust in the potential financial benefits of Galileo and motivated the private sector to reaffirm its participation in the funding of Galileo. The decision to move on to the development phase was taken at the Barcelona European Council in March 2002, confirming the concession model. The Council agreed on a cost share for the deployment phase 'of at most 1/3 for the Community budget and at least 2/3 for the private sector' (Barcelona European Council 2002: 16). After the Council finally reached a decision, the delay was prolonged by disagreements between the ESA Member States on the participation of national industries in Galileo. The ESA Council therefore needed until May 2003 to agree to the start of the development and validation phase (Court of Auditors 2009: 14).

In June 2003, the Commission and the ESA founded the Galileo Joint Undertaking (GJU) to manage the development and validation phase.[9] The GJU became operational in September 2003 and mainly served as a basis to coordinate the actions of the EU Commission and the ESA (Court of Auditors 2009: 14–15). The GJU had to select a preferred bidder for the Galileo concession from two bidding consortia by January 2005,[10] but did not make a decision. The GJU agreed to the request of the two bidding consortia in June 2005 to merge into one consortium and to make a joint offer. The negotiations for the Galileo concession and all technological activities of the development phase were stalled between July and December 2005 because of disagreement between some Member States about the merged consortium and about the location of ground components. An agreement was reached through mediation of a former EU Commissioner in December 2005 and negotiations with the consortium started in January 2006 (Court of Auditors 2009: 17). The Commission released a new timetable for Galileo in June 2006. The development and validation phase was extended to the beginning of 2009, the deployment phase was envisaged for 2009–2010, three years later than initially expected, and the commercial operation phase was planned to start in 2010. The budget for the development and validation phase was extended to €1.5 billion.

In January 2007, the tasks of the GJU were taken over by the European GNSS Supervisory Authority (GSA) (Commission 2006: 10–11). The GSA was installed in July 2004 to represent public interests and as the regulating entity for the European GNSS programmes during the deployment and operational phases of Galileo (Council 2004b: 3). In a Communication issued in May 2007, the Commission admitted to a five-year delay with Galileo, confirming that EGNOS had 'substantial cost overruns'. It advised the Council to end the PPP negotiations (Commission 2007: 4–6). In 2000, the Commission had estimated the overall costs of Galileo at a total of around €3.3 billion (Commission 2000: 26). In 2007, the Commission updated the overall costs to around €5.6 billion (Court of Auditors 2009: 19). The Commission (2007: 5) acknowledged misjudgement regarding risk allocation, the timing, budget and design of the programme and recognised failures in public and private governance.

According to a report issued in 2009 by the European Court of Auditors on the management of the Galileo development and validation phase, the Galileo PPP was 'inadequately prepared and conceived. As a result, the GJU was required to negotiate a PPP which was unrealistic' (Court of Auditors 2009: 22). The Court of Auditors stated that the Commission did not consider a number of best practices for PPPs: it failed to provide proper preparation, sufficient time, appropriate management resources, the maintenance of competition and regular reviews of the project. The Court of Auditors also pointed out that the Galileo PPP was different from other existing PPPs, especially given the high technological risk and because of hard-to-predict revenues given the fact that the open signal of the positioning system would be available free of charge (Court of Auditors 2009: 22–24). The Court of Auditors came to the overall conclusion that Galileo was not properly governed by the Commission as the roles had not been divided clearly enough between the Member States of the EU and the ESA, the Commission, and between the GJU and the ESA. It concluded that 'the Commission did not provide adequate leadership in developing and managing Galileo' (Court of Auditors 2009: 39). A member of the GNSS panel of the Council GSA pointed out in an interview that:

[a] PPP did not well match with the shape of Galileo. Maybe it was over-ambitious to adopt this approach for such a complex project. There is no similar PPP in space. The second reason could be that the PPP approach did not work for Galileo simply because it was not conceived for the right system or for the right phase of the system. Maybe it was over-ambitious, because to transfer the risk of a market which does not exist today at all was maybe simply too much.

(Personal interview, 27 June 2012)

Conclusions

Two distinct frames appeared during the Galileo PPP process. First, the negotiations were determined by a public value frame emphasising the advantages of the PPP concept for public infrastructure projects in the EU. The central terms of this frame are: value for money; claimed gains in efficiency enabled by the participation of the private sector; the advantage to the public sector of keeping PPP projects off the balance sheet; the ability of Galileo to generate profit; and the expected contribution of private finance to the programme. Second, a security frame focused on the independence of the EU and its Member States from existing satellite navigation systems, in particular from the dominant US GPS system, and on the potential military use of the Galileo PRS.

An analysis of the negotiations for the Galileo PPPs shows that the Commission underestimated the self-interest of the private sector and was too optimistic about the promises of the PPP concept. What seems clear is that the Commission lacked the necessary experience and know-how to lead the negotiations for the complex Galileo PPP under such difficult conditions. The DBFO concession model was not suitable for the Galileo project. A technically complex and politically sensitive project such as Galileo should be under public control from the outset, as was the case when the negotiations failed in 2007. The experience of Galileo has shown that the PPP concept, especially in a complex form such as a DBFO concession, might be more suitable for less demanding sub-contracts.

Other reasons for the failure were incorrect estimates of the overall costs and of the value for money of the project, which were affected by long delays in the political decision-making process. The situation was very difficult from the start, given that the Galileo project concerned so many national economic and security interests, greatly complicating the decision-making process and causing substantial delays in the negotiations. Delays in the political decision-making process, such as the postponed Council decision in 2000, made it harder for the private sector to accept the already high risks. The delays decreased the calculated revenues for the private sector, which, in turn, affected the ability and motivation of the private sector to accept the contract requirements. Finally, the private sector refused to bear the additional risks and the Council terminated the negotiations. After the termination of the PPP negotiations, Galileo was transformed into a public project entirely controlled and financed by the public sector. Financing for Galileo will be secured over the next six years through the Multiannual Financial Framework 2014–2020 of the EU, reserving €6.3 billion for the space programmes Galileo and EGNOS (Council 2013: 9).

The analysis of the negotiations for the Galileo PPP shows that the interest of the public and private actors to achieve individual aims, but also the interest to find a solution for a common problem, are of equal importance in the success of a PPP. It can be assumed that the definition of realistic goals and the prevention of considerable delays in the political decision-making process are key elements for the success of future negotiations for EU PPPs in the space sector, especially, if the new projects are as ambitious as Galileo. Only in this way will the private sector be ready to bear the high risks of such a project.

Notes

1 Value for money indicates substantial benefits for customers and taxpayers in terms of return on investment.
2 The European Commission defines e-Government as 'using the tools and systems made possible by Information and Communication Technologies (ICTs) to provide better public services to citizens and businesses' (European Commission Information Society 2010).
3 Before the financial crisis began in 2007, there was a constant growth of PPPs in the EU, but the crisis shook confidence in long-term and leverage funding and almost stopped the development of the European PPP market (DLA Piper/European PPP Expertise Centre 2009: 5). On the other hand, the fiscal problems of most EU Member States led to decreasing public long-term investment in infrastructure and supported the participation of private resources and know-how in public infrastructure projects (Mörth 2008: 40–41).

4 Concerning the weaknesses of the PPP concept, the Commission only refers to 'challenges' for PPPs because of the financial crisis, ignoring the weaknesses of the concept identified by critical studies on PPPs (Commission 2009: 10).
5 Transnational relations are defined by Keohane and Nye as 'regular interactions across national boundaries when at least one actor is a nonstate agent' (Keohane and Nye 1971: xii).
6 The development of Galileo was divided into four phases: technical definition; development and validation; deployment; and commercial exploitation (Commission 2000: 18–19).
7 New EU law regarding concessions was implied by the Commission one year after the report with Utilities Directive 2004/17/EC and the Public Sector Directive 2004/18/EC.
8 The eight objecting delegations were Austria, Denmark, the Netherlands, Sweden and the UK (Laeken European Council 2001).
9 The GJU was installed by Council Regulation EC 876/2002 in May 2002.
10 Eurely and iNavsat. A third consortium cancelled its application in 2004.

References

Barcelona European Council (2002) *Barcelona European Council 15 and 16 March 2002. Presidency Conclusions*. Report No. SN 100/1/02 REV 1. Available from: www. ael.pltt.edu/ld/eprlnt/43345. Last accessed 16 May 2013.

Baumgartner, F. and Mahoney, C. (2008) 'Forum section: the two faces of framing: individual-level framing and collective issue definition in the European Union'. *European Union Politics*, 9(3): 435–449.

Bertrán, X. and Vidal, A. (2005) 'The implementation of a public-private partnership for Galileo. Comparison of Galileo and Skynet 5 with other projects'. *Online Journal of Space Communication*, Issue 9. Available from: http://spacejournal.ohio.edu/issue9/pdf/Implement_Public-Private.pdf.

Bochinger, S., Wheeler, J., Dewar, J. and Franzolin, A. (2009) 'Applicability of public-private-partnerships in next generation Satcom systems'. Paper presented at the conference The Future of Public-Private-Partnerships in Satellite Communications, Vienna, 31 March 2009. Available from: www.espi.or.at/images/stories/dokumente/Presentations 2009/ppp_study_presentation_rev.pdf.

Bovaird, T. (2004) 'Public-private partnerships: from contested concepts to prevalent practice'. *International Review of Administrative Science*, 70(2): 199–215.

Commission (1998) *Communication from the Commission to the Council and the European Parliament: Towards a Trans-European Positioning and Navigation Network: Including a European Strategy for Global Navigation Satellite Systems (GNSS)*. COM(1998) 29 final, 21 January 1998.

Commission (1999) *Communication from the Commission: Galileo – Involving Europe in a New Generation of Satellite Navigation Services*. COM(1999) 54 final, 10 February 1999.

Commission (2000) *Commission Communication to the European Parliament and the Council: On Galileo*. COM(2000) 750 final, 22 November 2000.

Commission (2003) *High Level Group on the Trans-European Transport Network – Report*, 27 June 2003. Available from: http://ec.europa.eu/ten/transport/revision/hlg/2003_report_kvm_en.pdf.

Commission (2004) *Resource Book on PPP Case Studies, Directorate-General Regional Policy*, June 2004. Available from: http://ec.europa.eu/regional_policy/sources/docgener/guides/pppresourcebook.pdf.

Commission (2006) *Communication from the Commission to the European Parliament and the Council: Taking Stock of the GALILEO Programme.* COM(2006) 272 final, 7 June 2006.

Commission (2007) *Communication from the Commission to the European Parliament, the Council, the European Economic and Social Committee and the Committee of the Regions: Galileo at a Cross-road: The Implementation of the European GNSS Programmes.* COM(2007) 261 final, SEC(2007) 624, 16 May 2007.

Commission (2009) *Mobilising Private and Public Investment for Recovery and Long-term Structural Change: Developing Public Private Partnerships.* COM(2009) 615 final, 19 November 2009.

Commission (2011a) *Connecting Europe: The New EU Core Transport Network.* MEMO/11/706, 19 October 2011.

Commission (2011b) *Report from the Commission to the European Parliament and the Council: Mid-term Review of the European Satellite Radio Navigation Programmes.* COM(2011) 5 final, 18 January 2011.

Council (1999) *Council Resolution on the Involvement of Europe in a New Generation of Satellite Navigation Services – Galileo – Definition Phase Member of the GNSS Panel.* 1999/C 221/01, 19 July1999. Available from: http://eur-lex.europa.eu/LexUriServ/LexUriServ.do?uri=OJ:C:1999:221:0001:0003:EN:PDF.

Council (2001a) *Council Resolution on Galileo.* 2001/C 157/01, 5 April 2001. Available from: http://eur-lex.europa.eu/legal-content/EN/TXT/PDF/?uri=CELEX:52001XG053 0%2801%29&from=EN.

Council (2001b) *Written Question by Jacqueline Foster (PPE-DE) to the Council.* P-1502/01, 21 May 2001. Available from: www.europarl.europa.eu/sides/getDoc. do?pubRef=-//EP//TEXT+WQ+P-2001-1502+0+DOC+XML+V0//EN.

Council (2004a) *2629th Council Meeting – Transport, Telecommunications and Energy. Press Release, 9–10 October 2004.* C/04/345, 15472/04 (Presse 345). Available from: http://europa.eu/rapid/press-release_PRES-04-345_en.pdf.

Council (2004b) Council Regulation on the Establishment of Structures for the Management of the European Satellite Radio-navigation Programmes. 1321/2004, 12 July 2004. Available from: http://eur-lex.europa.eu/LexUriServ/LexUriServ.do?uri=OJ:L:2 004:246:0001:0009:EN:PDF.

Council (2013) *Conclusions (Multiannual Financial Framework).* EUCO 37/13, 8 February 2013. Available from: http://register.consilium.europa.eu/doc/srv?l=EN&f=ST%20 37%202013%20INIT.

Court of Auditors (2009) *The Management of the Galileo Programme's Development and Validation Phase.* Special Report 7/2009, 7 July 2009.

De Palma, A., Leruth, L. and Prunier, G. (2009) *Towards a Principal-Agent Based Typology of Risks in Public-Private Partnerships.* IMF Working Paper, WP/09/177. Washington, DC: International Monetary Fund.

DLA Piper/European PPP Expertise Centre (2009) *European PPP Report 2009.* Available from: www.eib.org/epec/resources/dla-european-ppp-report-2009.pdf. Last accessed 16 May 2015.

European Commission Information Society (2010) €90 Million Available for Research in Future Internet to Make Europe's Systems Smart and Efficient. Available from: http://ec.europa.eu/information_society/newsroom/cf/item-detail-dae.cfm?item_id=6021.

European Investment Bank (EIB) (2009) *EIB Financing of the Trans-European Networks.* Sectoral brochure QH-X1–06–058-EN-C. Brussels: European Investment Bank.

European Space Agency (2010) *Galileo Services.* Available from: www.esa.int/esaNA/SEMTHVXEM4E_galileo_0.html. Last accessed 16 May 2015.

Flinders, M. (2005) 'The politics of public-private partnerships'. *British Journal of Politics and International Relations,* 7(2): 215–239.

Grimsey, D. and Lewis, M. (2004) *Public Private Partnerships – the Worldwide Revolution in Infrastructure Provision and Project Finance.* Cheltenham/Northampton MA: Edward Elgar.

Hodge, G. and Greve, C. (2009) 'PPPs: the passage of time permits a sober reflection'. *Economic Affairs,* 29(1): 33–39.

Iossa, E., Spagnolo, G. and Vellez, M. (2007) *Contract Design in Public-Private Partnerships.* World Bank Report, Final Version, September 2007. New York, NY: World Bank.

Kappeler, A. and Nemoz, M. (2010) *Public-Private Partnerships in Europe – Before and During the Recent Financial Crisis.* Economic and Financial Report 2010/04, July 2010. Brussels: European Investment Bank.

Keohane, R. and Nye, J. (1971) *Transnational Relations and World Politics.* Boston, MA: Harvard University Press.

Laeken European Council (2001) 2393 *Council Meeting ECOFIN, Brussels, 4 December 2001.* Press Release 14627/01 (Presse 446).

Lembke, J. (2001) *The Politics of Galileo.* European Policy Paper No. 7, April 2001. Pittsburgh, PA: European Union Center, Center for West European Studies, University Center for International Studies, University of Pittsburgh.

Mörth, U. (2008) *European Public-Private Collaboration – A Choice Between Efficiency and Democratic Accountability?* Cheltenham: Edward Elgar.

Nardon, L. and Venet, C. (2010) *Galileo: the Long Road to European Autonomy. Actuelles de l'Ifri.* Available from: www.ifri.org.

PriceWaterhouseCoopers (2001) *Inception Study to Support the Development of a Business Plan for the GALILEO Programme.* TREN/B5/23–2001, 20 November 2001.

PriceWaterhouseCoopers (2005) *Delivering the PPP Promise – A Review of PPP Issues and Activity.* Available from: www.pwc.com/gx/en/government-infrastructure/pdf/promisereport.pdf. Last accessed 16 May 2015.

Renda, A. and Schrefler, L. (2006) *Public–private Partnerships. Models and Trends in the European Union.* DG Internal Policies of the Union, Directorate A, Economic and Scientific Policy, IP/A/IMCO/SC/2005–161. Available from: www.europarl.europa.eu/.../IPOL-IMCO_NT(2006)369859_EN.pdf. Last accessed 16 May 2015.

Riccardi, F. (2007) *The Galileo Project: What It Means, Its Goals, Mistakes, Questions (Sometimes Unanswered), Real Problems and Prospects for Relaunch.* Agence Europe, 14 June 2007.

Schäferhoff, M., Campe, S. and Kaan, C. (2009) 'Transnational public-private partnerships in international relations: making sense of concepts, research frameworks, and results'. *International Studies Review,* 11/2009, 451–474.

Weihe, G. (2008) 'Ordering disorder – on the perplexities of the partnership literature'. *The Australian Journal of Public Administration,* 67(4): 430–442.

14 Towards an EU industrial policy for the space sector[1]

Rik Hansen and Jan Wouters

Introduction

Recent years have seen the EU undertake or participate in an increasing number of space projects and ventures, thereby entering what had been for a long time the exclusive remit of the European Space Agency (ESA). It appears that, although teething troubles arguably continue to hamper the cooperation between the two organisations, a pragmatic approach has been adopted and the flagship programmes (Galileo and GMES) are being used to iron out any remaining problems in a bottom-up fashion. We therefore look at the lessons that have or should have been learnt from these initial experiences of the EU in the space arena. This chapter aims to do so with a particular focus on the field of industrial policy and procurement for two reasons. First, procurement – and more generally the respective funding approaches of both organisations – is one of the areas where the divide between the ESA and the EU has historically been most fundamental. Second, signs indicate that the EU is currently making an effort to develop an industrial policy that is suitably equipped for use in the very specific economic and political circumstances presented by the space sector, thereby encountering all sorts of obstacles of a legal and political nature. This chapter looks specifically at the experiences in the Galileo satellite navigation programme and attempts to decipher the sometimes equivocal messages the EU has given on the subject of industrial policy. To do so, a first section will examine the relationship between the EU and the ESA, looking first at the general picture, at the institutional aspects of the organisations' relationship and, finally, at the core problem at hand, their respective approaches to industrial policy. After this theoretical analysis, a second section looks at the current state of the Galileo project. It presents and examines the industrial policy choices made in the execution of Galileo, their effects and their reception by various space stakeholders. A third section then attempts to look ahead into the future, drawing what conclusions it can from the way industrial policy for space has been framed in recent EU and Member State discourse. Fourth, we conclude that a space-specific industrial policy is required to create a dependable environment for the space industry, to create a solid industrial base and thereby to enable the EU to play its self-proclaimed role as a space power.

The EU and the ESA

General

Without wanting to duplicate the extensive body of literature that has dealt with the intricacies of the institutional arrangements between the EU and the ESA (Schmidt-Tedd 2001; Wittig 2004; Reuter 2004; Hobe *et al.* 2006, 2009), it is worth briefly summarising the basis of their relationship.[2]

Of the two organisations, the EU is by far the larger, counting 28 Member States and over 500 million people who can call themselves citizens of the EU. This language by itself indicates that the EU has far surpassed its initial goal of creating economic disincentives to waging another war by economically intertwining the coal and steel markets of the major powers in Europe. Instead, it has become a political and social project of true unification with far-reaching economic, social, scientific, environmental, judiciary and even military consequences. In spite of it slowly becoming a political entity, complete with its own constituency, its underpinnings remain economic in nature, with the internal market and its four freedoms at its heart. A series of Treaties and Treaty modifications have created an autonomous, supranational legal order with considerable transfer of sovereignty from the Member States to the EU. In 2013, the EU's total budget was €142.6 billion (Europa 2014).

Compared with this amalgamation of objectives and competences embodied by the EU, the ESA's purpose appears to be simple. Its Convention, signed in 1975 by the combined Member States[3] of its predecessor organisations (ESRO and ELDO), defines the organisation's vocation as:

> to provide for and to promote, for exclusively peaceful purposes, cooperation among European States in space research and technology and their space applications, with a view to their being used for scientific purposes and for operational space applications systems.
>
> (ESA 1975: Article II)

To pursue this aim, the ESA was designed as a unique example of the flexibility afforded by public international law. Its activities fall into two categories: mandatory and optional programmes, with all Member States contributing to mandatory programmes in proportion to their GDP and the optional programmes being decided on an 'à la carte' basis.[4] In terms of its budget, the ESA is dwarfed by the EU, with a budget of €4.1 billion in 2014 (ESA 2014a, 2014b).

The approach of the two organisations to space is fundamentally different. The EU turns to space technology and space applications from an instrumental angle, appreciating them primarily for the potential they hold to further its core policies, but presently lacking the technical and organisational capability to harness it. For the ESA, the design, development and management of space projects constitute its core activities. An alliance therefore seemed logical. What was, however, also logical, is that such an alliance was bound to encounter difficulties because the

EU, with its utilitarian approach to space and its financial setup that implied funding space activities from the general EU budget, decides on programmes with a view to tangible, quantifiable and societally valuable outputs. The ESA, in contrast, decides whether to undertake programmes and how to fund them on a case-by-case basis at its Ministerial Council meetings that take place roughly every three years. For the ESA, each programme is an end in itself, with a separate inter-governmental agreement – albeit within the framework provided by the ESA Convention (ESA 1975) – detailing the programme definition, funding and industrial policy arrangements (Mazurelle *et al.* 2009, 11–14). One of the main motivations for Member States to sign up for an optional ESA programme lies precisely in the ESA's very specific industrial return rules, which make participating in a programme worthwhile – or at least defendable – on the basis of industrial return alone, regardless of further economic or societal benefits. Faced with these diverging motivations behind their interest in space applications, it was clear that significant ground needed to be covered to enable both organisations to cooperate harmoniously.

Institutional aspects

The only general agreement currently linking the EU to the ESA is the Framework Agreement signed in November 2003, which entered into force in May 2004 (Council 2004). Rather than wait for the outcome of a seemingly endless Treaty reform process, the EU and the ESA preferred to forge ahead and take some much-needed steps to accommodate their de facto cooperation in this Framework Agreement. It is worth noting that, from the EU's side, this agreement was concluded on the basis of then Articles 170 and 300(2) of the Treaty establishing the European Community (TEC), which invested the Community with the power to conclude international agreements in the fields of research and technological development (*Official Journal* 2002). As neither party had expressed its written intention to terminate this before the date set forward in the Agreement,[5] it has recently been prolonged for a third four-year period, meaning it will remain in force until at least 2016. Its main features consist of setting up the so-called Space Councils – that is, the periodic joint and concomitant meetings of the Competitiveness Council from the EU and the ESA Council at Ministerial Level, assisted in its functioning by a joint secretariat (Framework Agreement: Article 8). Since the entry into force of the Framework Agreement, eight such Space Council sessions have taken place. We will examine the contribution of these Council meetings to the progressive development of – at the very least – a joint EU–ESA discourse on industrial policy in some detail later in this chapter. The Framework Agreement provides no clear guideline on the subject, apart from the mention in Article 5.2 that an industrial policy scheme is one of the elements that should be agreed in case a joint initiative is undertaken that needs definition by specific agreements between parties (Framework Agreement: Articles 5.2). However, the same Article is quick to pre-empt any mixing of funds between parties, adding that '[u]nder no circumstances shall the European

Community be bound to apply the rule of "geographical distribution" contained in the ESA Convention and specially in Annex V thereto'. Institutionally, much remains to be settled between the EU and the ESA, the Framework Agreement arguably having outlived its time as a provisional measure to facilitate cooperation on an ad hoc basis. This section has, of course, ignored the various delegation agreements that have been concluded between both parties, in which the EU has entrusted the ESA with the technological development, procurement and construction of various parts of the Galileo and Copernicus programmes – a deliberate choice as they lack a sufficiently general character to be extrapolated to the whole of the organisations' relationship. We will return later to the indications for the future evolution of the EU–ESA relationship that have surfaced since the entry into force of the Lisbon Treaty.

Industrial policy and procurement

The Framework Agreement's strongly worded exclusion of ever applying a geographical distribution scheme – which is a cornerstone of the ESA's industrial policy and funding system – to EU funds may serve as an initial indication that industrial policy is an area of intense disagreement between the two organisations. First, however, we examine why specific industrial policy measures might be needed when commissioning space projects and what they may hope to achieve.

There are several reasons to treat the space sector differently from other sectors from an industrial policy point of view, most, but not all, of which are economic in nature.[6] Most of these specificities are jointly and unequivocally acknowledged by all the relevant space actors in Europe (Council 2010: 4; Commission 2013a: 5–10). Among the economic specificities, those cited most often are the high-technology content of space projects and space assets, the high cost of access to space and the relatively small size of the European (institutional) market (Hobe *et al.* 2011: 198). The high-technology content is reflected in high uncertainty in the early stages of development, in long development cycles and in the high initial capital requirements to be able to even start such a development cycle (Hobe *et al.* 2011). In turn, these factors combine to create formidable barriers to entry for both public and private actors wishing to undertake a space programme and for companies wishing to participate in the execution of one, especially at the prime contractor level (Hobe *et al.* 2011). The high cost of access to space means that space assets have to function autonomously for their entire lifetime, as on-orbit servicing opportunities are scarce to non-existent (Hobe *et al.* 2011).

Developing and launching space projects is normally a one-shot operation. The factors discussed here imply that important economies of scale can be achieved – although frequently they are not – and that large companies and institutional actors are better equipped to undertake these programmes. The European market – especially the institutional market – is relatively small in size, which implies that not all potential economies of scale are realised and space companies are dependent on the commercial market to a higher degree than else-where in the world.[7] One final consideration – non-economic in nature, but with significant economic consequences – is that the possession of, or capability to build and launch, space assets presents an important strategic interest to institu-tional actors such as national governments or the EU, not least because space assets are dual-use by nature (OECD 2005: 113, cited in Hobe *et al.* 2011: 201). This may engender policy objectives of strategic autonomy and non-dependence on third nations.

In addition to the very specific economic character of the market for space assets, the ESA came to space very early in the game, at a time when there was no well-structured industrial base for space in Europe. The legacy of its prede-cessor organisations (ESRO and ELDO) notwithstanding, the states negotiating the ESA Convention were conscious that the ESA would effectively create and shape a European space industry. Once it came into existence, the ESA would be the nascent industry's prime customer apart from the national space programmes. This consciousness is evidenced by the inclusion of industrial policy as one of four tools to pursue its purpose in Article II of the ESA Convention.

Article VII of the Convention provides more details on the objectives of the said industrial policy, stating that:

> [ESA's] industrial policy ... shall be designed in particular to:
>
> a. Meet the requirements of the European space programme ... in a cost-effective manner;
> b. *Improve the world-wide competitiveness of European industry ...*;
> c. Ensure that all Member States *participate in an equitable manner, having regard to their financial contribution ...*;
> d. Exploit the advantages of free competitive bidding in all cases, *except where this would be incompatible with other defined objectives of industrial policy.*
>
> Other objectives may be defined by the Council by a unanimous decision of all Member States ...' [emphasis added]

(ESA 1975: Article VII)

The ESA itself has always held that objectives (a–d) are of equal importance and that there is no hierarchy between them (Imbert and Grilli 2004; Petrou 2008). Plainly, however, the wording of objective (d) indicates otherwise and con-ditions the use of free competitive bidding on its compatibility with 'other defined objectives of industrial policy', thus de facto hierarchically subordinat-ing it (Hobe *et al.* 2011: 53). A clearer formulation would therefore be that three equally weighted objectives are pursued: meeting the requirements of the European space programme in a cost-effective manner; improving the world-wide competitiveness of European industry; and ensuring equitable participation – relative to their respective financial contributions – of all Member States. The provision should add that, wherever possible, these objectives are pursued using free competitive bidding.

The objectives of its industrial policy having been thus defined, Annex V to the ESA Convention further elaborates on the tools and techniques available. This is where public procurement rules and procedures enter the picture. First, Article II of Annex V contains an unambiguous Member State preference clause, providing that the ESA contracts should preferably be carried out by the industries of the ESA Member States and that ESA contracts in optional programmes should preferably be awarded to industries in those states participating in that particular programme. Second, Annex V establishes the geographical return system whereby the share of each Member State in the sum total of contracts awarded is divided by its share in financial contributions. The resulting figure is the return coefficient. When contracts are awarded in perfect proportionality to financial contributions, this coefficient is equal to 1, which is the target return coefficient for each Member State. In reality, such perfect equilibria are rare, and Annex V provides for a corrective mechanism whereby return statistics are periodically examined and deviations are corrected once they exceed a certain threshold (ESA Convention, Article IV *juncto* Annex V, Article V). At present, the minimum return threshold is set at 0.94 (Morel de Westgaver 2002, cited in Hobe *et al.* 2011: 73).

Further implementing instructions are contained in the ESA's Procurement Regulations and have undergone an important update, at once destined to make ESA procurement strategies more efficient and to bring them more in line with international procurement standards in terms of transparency. The decision to initiate the reform was taken at the 2008 ESA Council at Ministerial Level in The Hague. The most prominent measures were contained in an update of the Procurement Regulations that took place in 2009. This notably included the creation of an appeal procedure for interested parties who felt adversely affected by a procurement decision.

In conclusion, it is important to stress that the ESA's procurement system, characterised to the casual observer mainly by its geographical return mechanism – and, indeed, often reduced to this feature alone – is designed the way it is because it is a tool at the service of ESA's industrial policy objectives. Furthermore, it contributes greatly to the financial viability of the ESA model of intergovernmental cooperation as it provides Member States with a minimum guaranteed industrial return to justify their participation in an optional programme.

The EU also has an intricate procurement system, albeit one that is designed completely differently, the underlying purpose being not to create an industrial base in a specific sector (as is the case for the ESA), but rather to remove a myriad system of national preference and overt or covert protectionism in government procurement among EU Member States. Its basic provisions are contained in Article 322(1)(a) TFEU (ex. Article 279 TEC), which details the decision-making procedure for budgetary matters, and in Regulation 966/2012 (Council 2012) (the Financial Regulation, successor to Regulation 1605/2002;

Council 2002). Title V of this Regulation deals with procurement, with a scope covering all public contracts and framework contracts, but excluding grants, which are covered under a separate title (Financial Regulation: Article 101). There is no mention of using procurement to ends such as industrial policy in any of these basic legal documents; instead, the main principles governing procurement are those derived from primary EU law: transparency, proportionality, equal treatment and non-discrimination (Financial Regulation: Article 89, sub 1; Hobe *et al.* 2011: 94). Furthermore, the principle of sound financial management must be observed in all budgetary matters (TFEU: Article 319; Hobe *et al.* 2011: 94).

It has become apparent that, whereas both the EU and the ESA can and do procure in the space sector, they do so according to two very different sets of rules. This difference is a result of their fundamentally different approaches to procurement: the ESA sees and uses procurement as a tool in the implementation of its industrial policy, whereas the EU's procurement rules are targeted at ensuring the sound and equitable management of the EU budget and supporting the internal market project. The next section examines to what extent these differences have influenced procurement in the Galileo project.

Industrial policy in the Galileo project

Galileo history

Originally initiated in 1998, the European satellite navigation programme Galileo has had a turbulent history. In 2007, commentators called for the Galileo project to be scrapped in no uncertain terms, calling it 'a political creation founded on national vanities rather than commercial logic' with '[a] multi-country structure [that] results in internal discord' (*The Economist* 2007). Although it is certain that Galileo's history has been one with as many troughs as peaks, it must be said that 2007, when the failure of the public–private partnership model that should have financed the technological development phase became apparent, was perhaps its darkest hour (see Chapters 13). In the time since that particular low point, the programme has been resurrected, fully funded by the EU this time, a tripartite governance scheme has been set up, contracts have been granted and four in-orbit validation satellites – which will continue to function in the final constellation – have been launched. A first positional fix using the signals transmitted by these satellites was achieved in March 2013; the full constellation is projected to be operational in 2020, with service availability increasing gradually up to that time (Commission 2013b, 2013c). The next section looks at the industrial policy and procurement mechanisms that were used in the procurement round for the full operational capability (FOC) phase of the project.

Industrial policy in Galileo procurement

Public procurement is one of the most potent instruments that can be used to pursue objectives of industrial policy. In designing a procurement plan for the FOC phase of Galileo, the EU was faced with a double problem: the Financial Regulation provided no suitable procurement instruments for such a specific programme, yet any disbursement of funds from the general budget – as is the case for Galileo's FOC phase[8] – would have to comply with it. A solution was found by adopting a separate implementing Regulation, Regulation 683/2008 (GNSS Implementing Regulation; Council and European Parliament 2008), to deal with governance and procurement for the European GNNS programmes Galileo and EGNOS. Although the Regulation lacks any mention of industrial policy by name, Article 17 sets out a number of objectives that are highly reminiscent of such a policy:

> Procurement principles related to the deployment phase of the Galileo programme
>
> 1   The Community's public procurement rules ... shall apply to the deployment phase of the Galileo programme without prejudice to measures required to protect the *essential interests of the security of the European Union* or public security or to comply with European Union export control requirements.
> 2   During the procurement, the following objectives shall be pursued:
>
>   (a) promoting the balanced participation of industry at all levels, including, in particular, SMEs, *across Member States*;
>   (b) avoiding possible *abuse of dominance* and avoiding long-term reliance on single suppliers;
>   (c) taking advantage of prior public sector investments and lessons learned, as well as industrial experience and competence, including that acquired in the definition and development and validation phases of the programmes, while ensuring that the rules on competitive tendering are not prejudiced.
>   (Regulation 683/2008: Article 17 [emphasis added])

A first observation is that the GNSS Implementing Regulation makes room for a number of considerations not included in the Financial Regulation, notably relating to the specific strategic interest that space technology and space assets represent and to the specific market conditions in the space market. Article 17 then continues with a number of project-specific procurement techniques, designing the Galileo FOC procurement in such a way as to achieve the stated objectives:

> (a) the procurement of the infrastructure shall be split into a set of six main work packages ...;

(b) … any one independent legal entity … may bid for the role of prime contractor for a maximum of two of the six main work packages;

(c) at least 40% of the aggregate value of the activities shall be subcontracted by competitive tendering […];

(d) dual sourcing shall be pursued wherever appropriate.…

(Regulation 683/2008: Article 17, sub 3)

Evaluation

With the contracts resulting from procurement according to this procedure having been signed between January 2010 and June 2011, and work on the goods and services procured now well underway, the time is ripe for an evaluation. First, we make some remarks specific to this round of procurement and, second, we look at the overall state of industrial policy in European space policy.

Commenting generally on the approach chosen by the EU for the procurement in the FOC phase of Galileo, it is immediately apparent that this is a one-off approach, designed specifically for a given programme under given market conditions. In particular, the determination of the number of work packages, the maximum number of work packages for which any one entity was allowed to tender and the 40 per cent subcontracting target seem intimately entangled with the specific circumstances of the Galileo project. Judicious though these choices may prove to be in this particular project, some concerns may be raised when it comes to upscaling them to a more general procurement regime for EU space projects. In addition, some of the objectives set forth here appear hard to quantify, verify and enforce, particularly the subcontracting target (which several stakeholders from the space industry thought would be hard to implement as it was formulated as a preference, instead of as a binding rule)[9] and the preference for dual sourcing 'wherever appropriate'. On a more fundamental note, the order of priorities in this procurement scheme seems somewhat skewed. By introducing an implementing Regulation and a number of objectives that closely resemble the use of procurement as an industrial policy measure, albeit not under that name, the EU has reversed the customary order of things. It has attempted to mitigate the consequences of applying its general procurement system to a type of project it was not designed to cater for by infusing it with non-procurement objectives. It has thereby arrived at what might be termed an EU law-compatible, but ESA-inspired, industrial policy that has at least some concern for the market conditions it faces and the market it shapes. Although this approach may be defendable in this particular case, it is no substitute for developing a sector-specific industrial policy with clearly formulated objectives and subsequently equipping it with suitable procurement tools. This conclusion is also reached by Hobe *et al.* (2011). In the next section, we examine the steps undertaken in this direction by the EU since the completion of this space procurement milestone.

Framing the EU's march towards a space industrial policy

European space governance is in a constant state of evolution. The legal provisions and governance arrangements under which the Galileo programme is taking shape are therefore arguably already outdated at a time when the first operational satellites have only just reached orbit. It is our opinion that the driver for future adaptations of the governance and procurement schemes for EU space activities is the much-debated Article 189 TFEU, which still raises questions as to its precise effects more than four years after its entry into force (Mazurelle *et al.* 2009: 23–33; Wouters 2009: 116–124). This section attempts to draw some preliminary conclusions on the basis of discourse in policy documents adopted in the meantime: the Resolutions of the Seventh and Eighth Space Councils, Commission Communications on space policy, on the relationship between the EU and the ESA, on industrial policy in general and, finally, the Commission Communication released in early 2013 on the Commission's views on a sector-specific industrial policy for the space sector. In our attempt to do so, we will borrow elements from political science, more specifically the theory of framing, as our backdrop (Surel, 2000, Stephenson 2012). The question we ask is whether the Commission and the Council have introduced new frames in their respective space discourse since the entry into force of the Lisbon Treaty and Article 189 TFEU and to what extent they signal new stages in the progression towards a sector-specific industrial policy for space. Although it is beyond the scope of this chapter to provide the definitive framing analysis of the EU's nascent space policy, we hope to show the richness and variety of frames employed by the policy-making actors in the specific context of industrial policy.

In the documents emerging from the Space Council, consisting of Member State delegations from the EU and ESA Member States, we find relatively little mention of industrial policy. The Resolution of the Seventh Space Council, the first meeting of the Space Council since the entry into force of the Lisbon Treaty, contains a single paragraph on industrial policy:

> CONSIDERS that industrial policy for space should take into account the specificities of the space sector and the interest of all Member States to invest in space assets, and aim at the following common objectives: support the European capability to conceive, develop, launch, operate and exploit space systems; strengthen the competitiveness of European industry for both its domestic and export markets; and promote competition and a balanced development and involvement of capacities within Europe.
>
> (Council 2010: 4)

The Eighth Space Council repeats this formulation verbatim, adding only that it is necessary 'to examine whether appropriate measures may be necessary at European and international level to guarantee the sustainability and economic development of space activities' (Council 2011: 10). Without clarifying who should take the initiative for drawing up an industrial policy for space, or what instruments or measures it should consist of, the Resolutions indicate at the very least the agreement between the EU and the ESA on the need for such an industrial policy, as well as on a number of its objectives. The frames we tentatively identify in these paragraphs are the following: (1) the space sector presents economic specificities that need to be taken into account; (2) investing in space activities should be in the interest of Member States individually; (3) Europe should invest in the development of a set of capabilities; (4) Europe should support its industry's competitiveness internally as well as externally; and (5) competition as well as balanced development and involvement across Europe should be pursued.

The Commission from its side, in a 2010 Communication on general industrial policy, already clearly moots the possibility of a sector-specific industrial policy for space, mentioning, in particular, the new competence conferred to the EU in the Lisbon Treaty (Commission 2010: 23). It stresses the economic specificity of the space sector (Commission 2010: 24) and sets forward a number of objectives to be pursued by a space industrial policy: fostering a solid and balanced industrial base; ensuring greater international competitiveness; ensuring non-dependence in strategic sectors; and developing downstream markets (Commission 2010). In conclusion, it states that the Commission will 'pursue a Space Industrial policy developed in close collaboration with the European Space Agency and Member States' (Commission 2010). It is important to note that, in this Communication, the Commission echoes four out of the five frames identified in the Space Council Resolutions, leaving out only the second frame on individual Member States' interests in investing in space.

The Commission Communication of 11 April 2011, building on the bricks laid by the Communication on industrial policy, contains a separate subsection on industrial policy. It declares space to be an integral part of the Europe 2020 strategy, citing its potential as a driving force for growth and innovation (Commission 2011b: 8). It notably reiterates the dependence of the space sector on public procurement and stresses the need for appropriate procurement instruments, including '[t]he option of adopting specific provisions under particular legislative acts'. In doing so, the Commission once more affirms all but one of the Space Council's frames for space industrial policy, remaining silent only on the individual Member State perspective. It adds two important frames: (6) that space is an integral part of EU innovation policy; and (7) that the EU benefits from an express mandate in Article 189 TFEU to draw up an industrial policy.

The Commission's focus returned to space policy during the second half of 2012 and the first half of 2013, when it shared its views on the medium- to long-term evolution of the relationship between the EU and the ESA (Commission 2012) and on a future sector-specific industrial policy for the space sector in a Communication on the subject (Commission 2013a). The latter document once again stresses the specific nature of the space sector, both from an economic and from a strategic perspective, warranting measures of industrial policy to remedy some of the market's shortcomings (Commission 2013a, 5–10). The goals for such a policy are by and large in line with earlier indications:

- establishing a coherent and stable regulatory framework;
- developing a balanced industrial base and ensuring a healthy degree of participation by SMEs;
- improving worldwide competitiveness through improved cost-efficiency;
- developing downstream markets;
- ensuring technological non-dependence and independent access to space.

(Commission 2013a: 4)

At the same time, the Commission goes to great lengths to show that, for the EU, space is a policy area that is wholly integrated with the rest of its competence catalogue. As such, the industrial policy measures that it has at its disposal are much broader than procurement alone, ranging from the promotion of innovation under the Europe 2020 growth strategy to the standardisation of technology, education, trade and the promotion of market integration (Commission 2013a: 11–21). It is quick to add that developing a balanced industrial base 'does not mean the equal spread of this niche industry all over Europe, but an industry that builds on competitive advantages of the whole supply chain' (Commission 2013a). Frame-wise, this is wholly in line with previous Commission instruments. Once more, the motif of individual Member State interests in contributing financially to space programmes is conspicuously absent.

From all these elements, it seems fair to conclude that the EU has learnt from the somewhat illogical industrial policy and procurement arrangements in the Galileo project and that a tailored space industrial policy is underway. Nevertheless, even if the Commission proposal is only an early indication, the end result should be expected to be far different from what has long been understood as space industrial policy under the ESA system. We have briefly examined the frames used by the various actors in the policy dialogue that has ensued the entry into force of Article 189 TFEU. It appears from the fact that the majority of these frames are shared between the Space Council and the Commission that there is at least some common ground between the EU and the ESA's Member States. At the same time, some disparity in the frames employed hints at points of discord. From the side of the Member States and the Space Council, a point not taken up by the Commission – and, we would guess, significantly so – is their assertion that a space industrial policy should ensure the alignment of individual Member States' interests in contributing financially to such a programme. The Commission, from its side, places more emphasis on the EU's legitimacy as a space actor and on the integration of space policy and space industrial policy in the broader ambit of the EU's policies.

Conclusion

For the EU, the role of industrial policy in the space sector is two-fold. At the most general level, Article 189 TFEU expressly recognises that space activities are a tool at the service of a general EU industrial policy. This is echoed by the

EU 2020 agenda and by the Commission Communication on industrial policy, both of which recognise the potential that the space sector holds as a driver for innovation and growth, while acknowledging the specific economic characteristics of the space market. At a lower level, industrial policy in the space sector also serves the purpose of creating a solid industrial base for European space activities with a sufficient measure of competitiveness, both in the internal market and in the world market. The latest Space Council Resolutions, the Commission Communications on industrial policy, on space strategy and on institutional relations between the EU and the ESA furthermore provide progressively more and clearer indications of the shape a sector-specific space industrial policy might take. Our framing analysis of the policy dialogue highlighted the fact that, in spite of diverging accents between the Commission and the respective Member States of the ESA and the EU, there is significant overlap in their opinions of what the goals of such a space industrial policy should be. On the basis of the latest Communication on space industrial policy, we hope that the EU will carry out its Article 189 TFEU mandate in full and that this process will result in a tailored industrial policy that will allow the European space sector to become the driver for growth and innovation that the EU trusts it can be.

Notes

1 This chapter is a comprehensively updated version of an article published previously as Hansen and Wouters (2012).
2 For an in-depth look at the history of EU–ESA relations, see Hobe *et al.* (2006) and Hobe *et al.* (2009).
3 minus Australia, which was an associate Member State of ELDO.
4 For a comprehensive account of the history of Europe in space and the ESA in particular, see Madders (1997).
5 One year before the end of each four-year period (Framework Agreement: Article 12).
6 This section is partly indebted to the SP4ESP Space Procurement for the European Space Policy research project carried out under FP7 funding (Grant Agreement No. 242286) and published as Hobe *et al.* (2011).
7 The total sales of European space companies in 2012 amounted to €6.5 billion, of which 53 per cent (€3.5 billion) was accounted for by public customers (ASD-Eurospace 2013).
8 The total amount allotted to Galileo under the 2007–2013 multiannual financial framework is €3.4 billion (Commission 2011a: 11); a further €6.3 billion (in 2011 prices) has been allotted under the 2014–2020 Financial Perspectives (Regulation 1312/2013: Article 17).
9 The stakeholder opinions represented here were in part gathered in interviews conducted for the SP4ESP Space Procurement for the European Space Policy research project carried out under FP7 funding (Grant Agreement No. 242286) and published as Hobe *et al.* (2011).

References

ASD-Eurospace (2013) *State of the European Space Industry in 2012 (Web Release), June 2013*. Available from: eurospace.org/Data/Sites/1/simpaper2013draftfinal.pdf. Last accessed: 12 February 2014.

Commission (2010) *Communication from the Commission to the European Parliament, the Council, the European Economic and Social Committee and the Committee of the Regions: An Integrated Industrial Policy for the Globalisation Era Putting Competitiveness and Sustainability at Centre Stage*. COM(2010) 614, 28 October 2010.

Commission (2011a) *Report from the Commission to the European Parliament and the Council: Mid-term Review of the European Satellite Radio Navigation Programmes*. COM(2011) 5 final, 18 January 2011.

Commission (2011b) *Communication from the Commission to the European Parliament, the Council, the European Economic and Social Committee and the Committee of the Regions: Towards a Space Strategy for the European Union that Benefits its Citizens*. COM(2011) 152, 11 April 2011.

Commission (2012) *Communication from the Commission: Establishing Appropriate Relations between the EU and the European Space Agency*. COM(2012) 671, 14 November 2012.

Commission (2013a) *Communication from the Commission: EU Space Industrial Policy. Releasing the Potential for Economic Growth in the Space Sector*. COM(2013) 108, 28 February 2013.

Commission (2013b) *First Steps of Galileo – European Satellite Navigation System Achieves its First Position Fix. Press Release 12 March 2013*. Available from: www. ec.europa.eu/enterprise/policies/satnav/galileo/files/2013-03-12_galileo_position_fix_march_2013_en.pdf. Last accessed 12 February 2014.

Commission (2013c) *Galileo, Europe's GPS, Opens Up Business Opportunities and Makes Life Easier for Citizens, Memo 24 July 2013*. Available from: www.europa.eu/rapid/press-release_MEMO-13-718_en.htm. Last accessed 17 May 2015.

Council (2002) Regulation No. 1605/2002 of 25 June 2002. *Official Journal*, 16 September, L 248, 1.

Council (2004) 2004/578/EC: Council Decision of 29 April 2004 on the Conclusion of the Framework Agreement between the European Community and the European Space Agency. *Official Journal*, 6 August, L 261, 63–68.

Council (2010) *Seventh Space Council Resolution: 'Global Challenges: Taking Full Benefit of European Space Systems'*. Council Document 16864/10, 26 November 2010.

Council (2011) Eighth Space Council: Orientations Concerning Added Value and Benefits of Space for the Security of European Citizens. Council Document 17828/1/11, Rev. 1, 6 December 2011.

Council (2012) Regulation No. 966/2012 of the European Parliament and of the Council of 25 October 2012. *Official Journal*, 26 October, L 298, 1.

Council and European Parliament (2008) Regulation No. 683/2008 of the European Parliament and of the Council of 9 July 2008. *Official Journal*, 24 July, L 196, 1.

Europa (2014) *European Union Budget 2014 in Figures*. Available from: www.ec.europa.eu/budget/figures/2014/2014_en.cfm. Last accessed: 12 February 2014.

European Space Agency (ESA) (1975) *Convention for the Establishment of a European Space Agency*. UNTS Vol. 1297 I-21524, 30 May. Paris: ESA.

European Space Agency (ESA) (2014a) *Procurement Regulations and Related Implementing Instructions, ESA/REG/001, rev. 3, Paris, 20 December 2012*. Available from: www.emits.sso.esa.int/emits-doc/e_support/GCE/esa_reg_001revision_3_final_dg.pdf. Last accessed: 12 February 2014.

European Space Agency (2014b) Welcome to ESA. Available from: www.esa.int/About_Us/Welcome_to_ESA/Funding. Last accessed 12 February 2014.

Hansen, R. and Wouters, J. (2012) 'Towards an EU industrial policy for the space sector – lessons from Galileo'. *Space Policy*, 28(2): 94–101.

Hobe, S., Kunzmann, K., Reuter, T. and Neumann, J. (2006) *Rechtliche Rahmenbedingungen einer zukünftigen kohärenten Struktur der europäischen Raumfahrt*. Münster: Lit.

Hobe, S., Heinrich, O., Kerner, I. and Froehlich, A. (2009) *Entwicklung der Europäischen Weltraumagentur als "implementing agency" der Europäischen Union: Rechtsrahmen und Anpassungserfordernisse*. Münster: Lit.

Hobe, S., Hofmannová, M. and Wouters, J. (eds) (2011) *A Coherent European Procurement Law and Policy for the Space Sector: Towards a Third Way*. Münster: Lit.

Imbert, P. and Grilli, G. (2004) 'La politique industrielle de l'ESA – le concept évolutif du "juste retour " '. *ESA Bulletin*, 78, 16–XX.

Madders, K. (1997) *A New Force at a New Frontier*. Cambridge: Cambridge University Press.

Mazurelle, F., Wouters, J. and Thiebaut, W. (2009) 'The evolution of European space governance: policy, legal and institutional implications'. *International Organizations Law Review*, 6, 1–35.

Official Journal (2002) Treaty Establishing the European Community, Consolidated Version. *Official Journal*, 24 December, C 325.

Petrou, I. (2008) 'The European Space Agency's procurement system: a critical assessment'. *Public Contract Law Journal*, 37(2), 146–177.

Reuter, T. (2004) 'The framework agreement between the European Space Agency and the European Community: a significant step forward?' *Zeitschrift für Luft- und Weltraumrecht*, 53, 56–65.

Schmidt-Tedd, B. (2001) 'Rechtlichte Implikationen der gemeinsamen ESA/EU-Raumfahrtstrategie'. *Zeitschrift für Luft- und Weltraumrecht*, 50, 202–216.

Stephenson, P. (2012) 'Talking space: the European Commission's changing frames in defining Galileo'. *Space Policy*, 28(2): 86–93.

Surel, Y. (2000) 'The role of cognitive and normative frames in policy-making'. *Journal of European Public Policy*, 7(4): 495–512.

The Economist (2007) 'European industrial policy: lost in space', 12 May 2007. Available from: www.economist.com/node/9150765. Last accessed 12 February 2014.

Wittig, S. (2004) 'Die neue Aufgabenverteilung zwischen ESA und EU in der Raumfahrt'. *Zeitschrift für Luft- und Weltraumrecht*, 53(3), 415–423.

Wouters, J. (2009) 'Space in the Treaty of Lisbon'. In: K.-U. Schrogl, C. Mathieu and N. Peter, (eds), *Yearbook on Space Policy 2007/2008: From Policies to Programmes*. Vienna: Springer, 116–124.

15 Role of Galileo satellite technology in maritime security, safety and environmental protection

Angela Carpenter

Introduction

This chapter discusses how the European Space Agency (ESA) and European space policy (ESP) have sought to make a case for the benefits and synergies that can be gained from working with what is traditionally considered an unrelated area – EU maritime policy and the marine environment. Since the establishment of the European Maritime Safety Agency (EMSA) in 2002 (Commission 2002), ESA and the ESP have provided support to enable EMSA to carry out its work. This chapter identifies how institutional actors have sought to promote satellite technology and the crucial role it can play in improving the security, safety and environmental protection of the EU's maritime regions.

The development of policies may be in response to the bias or opinions of the policy actor, individually or collectively, resulting in policy areas being transferred from the national to the supranational level (Dudley and Richardson 1999: 226). This might depend, for example: on whether that policy is one-dimensional or multi-dimensional and whether it is the result of individual or collective action to push forward that change (Baumgartner and Mahoney 2008: 435–439); on the role of the institutions and actors working towards institutional change and how such change meets the preference of the actors driving it forward (Kohler-Koch 2000: 515); and on how, despite the European Commission being viewed as a coherent and strategic actor, it is a complex organisation that has complex relationships with external actors in areas such as industry and defence (Mörth 2000: 173–189).

This chapter is set in the context of the hypothesis of Hörber (2012: 78) that Europe has developed beyond the guiding ideals of European integration for the purpose of peace and prosperity after the Second World War and has now entered a phase of consolidation and exploration. This idea was supported by Manners (2002: 240), who identified an evolution from post-war nationalism towards a Europe of pooled resources and common principles. Similarly, Vogler and Stephan (2007: 390) noted that the EU pursues collective action and advances its own regional integration to establish an identity for itself as an actor on the global stage.

Multilateral action and cooperation in the area of space policy would garner respect for Europe as a global partner, raise the EUs political standing in the world, improve EU economic competitiveness and also enhance its scientific reputation (Commission 2003a: 6–7).

The European Commission, in setting out preliminary requirements for an EU space policy, emphasised the vital role space can play in EU policies such as transport, environment and security, and the integration of space and terrestrial components in areas such as monitoring and communication (Commission 2005: 5). More recently, the European Commission set out that:

> Europe needs an effective space policy that will allow the EU to take the global lead in selected strategic policy areas. Space can provide the tools that address many of the global challenges that face society in the twenty-first century: challenges that Europe must take a lead in addressing.
>
> (Commission 2011a)

The ESP provides a clear example of the peaceful integration of European activities over the last few decades. It shows how the development and exploitation of technology is an area where ESA can contribute towards the prosperity of European Member States (Hörber 2012: 78), despite the economic problems and other regional and social disparities that still exist across Europe. For example, through cooperative activities between ESA and the EU, such as the creation of the Global Monitoring for Environment and Security initiative, the ESP can also be viewed as a move towards greater integration and consolidation through the development of a European identity using space policy (Hörber 2012: 78–79).

Complementing these studies, this chapter examines how the EU has, in recent years, sought to implement and apply aspects of space policy within the maritime policy arena (see, for example, Commission 2006: 31–33; Commission 2007: 5–7). By considering synergies between the two policy areas, it is apparent that they are both multi-dimensional, with many diverse aspects of space policy able to support and benefit marine policy. That support is occurring in areas including environmental protection, weather forecasting and route planning, and in supporting both national and supranational agencies in actions as diverse such as security, policing, fisheries and transport, and in different ways in different EU maritime regions. The European Commission notes, for example, that priority should be given to developing services that include: 'Ocean monitoring to improve understanding of climate change and to support sustainable management of resources e.g. fisheries [and that] Maritime transport requires adequate surveillance for increased safety and environmental protection' (Commission 2003a: 13).

Synergies between the space and maritime sectors

Under the guiding principle of peaceful cooperation between European states, the purpose of the 1975 ESA Convention was to promote space research and technology, with that technology to be used for scientific purposes and for operational space applications systems (ESA 2003: 10). The Director-General of ESA set out, in 2007, a Resolution on an ESP providing a common political framework for space activities in Europe that recognised the need for stronger cooperation between ESA and the EU to promote a unified approach to space, highlighting the significant contribution an ESP could make to an independent, secure and prosperous Europe (ESA 2007a: 9).

The European Commission has identified the importance of space activities and applications for the growth and development of society by promoting the well-being of its citizens, providing economic benefits and by cementing the EU's position as a major player on the international stage (Commission 2011b: 2). In space policy, a strong European/EU political identity is particularly important where the EU represents its Member States on the international stage. The European Commission also emphasised the need for the EU to increase its participation in the governance system and activities of the United Nations, in areas such as supporting the United Nations in combating terrorism, drug trafficking and organised crime (Commission 2003b: 7–8). It can therefore be considered that these are all areas where there is the potential for synergies to occur between space and maritime policy actions.

The application of space research and technology has been beneficial to the EU and its wider marine environment for many years in the areas of marine environmental protection and maritime safety and security. ESA has been actively involved in the maritime sector for over 25 years (ESA 2007b) and access to a wide range of natural and human activity data to enable strategic decision-making on maritime policy was one area identified by the European Commission as a key component of its integrated maritime policy (Commission 2007: 6).

Topics for shared action and collaborative projects between EU institutions using earth observation in the area of marine science and technologies have included coastal zone research (the exploration and exploitation of resources), ocean currents and water circulation studies, and polar (ice thickness) research (Contzen and Ghazi 1994: 102).

Satellite data for earth observation, with tools such as remote sensing satellites, have come to play a significant role in monitoring a state's compliance with its international obligations under environmental agreements and for conducting environmental research (Peter 2004: 189–190).

ESA, the ESP and EMSA

EMSA was established in 2002 following the sinking of the Merchant Vessel *Erika* in December 1999.[1] It is the EU Agency with responsibility for ensuring uniform and effective maritime safety and the prevention of pollution from ships operating in EU waters (Commission 2002). EMSA was tasked with providing objective, reliable and comparable information and data in those areas that would enable Member States to take steps to improve safety and prevent pollution by collecting, recording and evaluating data on maritime safety, maritime traffic and both accidental and deliberate pollution (Commission 2002: Article 2, Task (f): 4).

Although the use of satellite imagery and earth observation to obtain such data was not stated explicitly in its establishing Regulation, it appears implicit that EMSA should obtain data from any source possible and that the work of ESA and the ESP will assist EMSA in fulfilling its tasks. However, even prior to the establishment of EMSA, satellites were used for earth observation in different maritime policy areas, including oil slick detection, with ESA (2007c) indicating that, in 2002, the newly launched ENVISAT was able to acquire images of the *Prestige* oil spill off Spain (Figure 15.1).

Advanced synthetic aperture radar imagery from ENVISAT has been used in conjunction with other systems to identify ice coverage, enabling ship operators

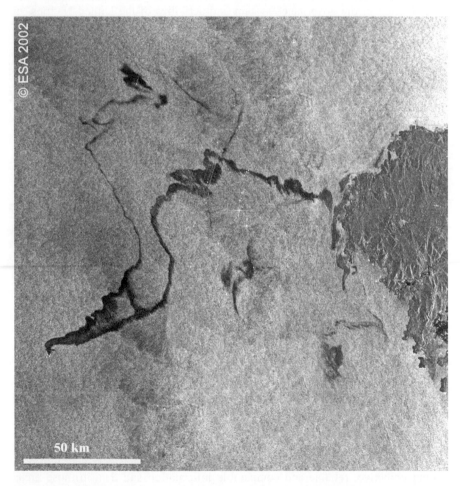

Figure 15.1 ENVISAT image of the *Prestige* oil slick, November 2002.

to avoid hazards such as icebergs or sea ice threats to commercial shipping in the Baltic Sea (ESA 2004a). Figure 15.2 shows an ENVISAT image of ice, up to 80 cm thick, covering parts of the Gulfs of Bothnia, Finland and Riga. This has had implications for both safety and search and rescue activities in the event of accidents, while accurate route planning data have enabled ships to take advantage of ocean currents or avoid severe weather, which can result in faster voyages and a reduction in fuel consumption, potentially reducing greenhouse gas emissions from ships (ESA 2004b).

The EMSA website, in discussing the practical application of satellite data[2] gives an example where a ship carrying dangerous or polluting goods is involved in a collision. With the availability of up-to-date information on the locations of all vessels using monitoring systems in the area, a hazard warning can be sent to all those vessels to avoid the specific area and to warn other Member States along the ship's route of the potential risk from a spill.

Figure 15.2 ENVISAT image of Baltic ice coverage, April 2011.

Automatic identification systems for vessel tracking and identification can support search and rescue activities and can also play a significant role in ensuring the security of the EU's borders with maritime security. ESA (2008) noted that this is an area where it has responded to requirements for improved vessel identification and location tracking, not just so that EU coastal states can better control and manage access to their territorial waters and exclusive economic zones for fisheries, but also to counter illegal activities such as the trafficking of people, drugs or weapons. ESA has therefore been involved in the use of operational satellite technologies for vessel tracking as part of the overarching International Maritime Organization's Long Range Identification and Tracking of Ships system, part of a global system administered by the International Maritime Organization under the aegis of the Safety of Life at Sea Convention 1974,[3] the EU component of which is managed by EMSA.

Two further operational tasks and services of EMSA[4] that use satellite technologies are: satellite oil spill monitoring through CleanSeaNet; and vessel traffic monitoring in EU waters under SafeSeaNet. CleanSeaNet is a system that uses radar satellite images to identify oil pollution on the sea surface, monitor accidental pollution during emergencies and contribute to the identification of polluters. Member States are alerted to possible oil spills and, if a spill is confirmed, the polluting vessel may potentially be identified by vessel tracking through the SafeSeaNet system. As a result, this activity can enhance maritime safety, port and maritime security, marine environmental protection and also improve the efficiency of maritime traffic and maritime transport through a centralised maritime data exchange platform that monitors ships operating in EU waters.

Satellite systems and maritime safety and security

There are many different actors in EU Member States and EU agencies, including general law enforcement, customs, marine environment, maritime safety and security, defence, fisheries control and border control (Commission DG-MAF 2010) with an interest in, or responsibility for, aspects of maritime safety and security. In each of these areas there is a range of (sometimes overlapping) issues to be dealt with.

The automatic identification vessel tracking and identification systems, using technology deployed on satellites by a range of different companies and government programmes, can assist civilian and military agencies, such as EMSA, to daily monitor the position of up to 17,000 ships in and around EU waters (EMSA 2012a). SafeSeaNet can identify the current position of all or a single ship in and around EU waters, display specific vessel types and show their historical positions (EMSA 2012b). Different types of vessel are required to carry Automatic Identification System transmitters on board and failure to do so may be viewed as a flag for investigating these ships before they enter EU waters or ports. Automatic identification system data could also support monitoring of compliance with regulations or enforcement operations and aid in the identification of illegal fisheries and fish landings (Commission DG-MA 2010 Annex: 23).

Under the SafeSeaNet system, information is also held about ships carrying hazardous materials, dangerous and polluting goods, or substances that can cause environmental damage or pose a hazard in the event of an accident. If hazardous materials are spilt, the system can enable search and rescue agencies to identify the potential hazards faced during rescue operations. In addition, it allows pollution monitoring and prevention agencies to monitor high-risk vessels sailing in EU waters (Carpenter 2012: 261–262). EU Long Range Identification and Tracking of Ships, SafeSeaNet and Automatic Identification System data may be used to identify nearby vessels and request that they move to a position to render assistance to a vessel in distress.

Satellite positioning has been undertaken for over 20 years through the Global Navigation Satellite System, which has been in operation since the early 1990s. Originally using the US global positioning system and the Russian GLONASS system, both under military control, Strodl *et al.* (2004) note that the introduction of the European Galileo system, which is under civil control, offers a 'certified Safety of Life Service (SoL) [which can be] used for all applications which require a certain level of reliability and integrity of positioning', particularly in relation to 'transport applications, where lives are at stake if the performance of the navigation system is degraded' Strodl *et al.* (2004: 8). In addition, they note that Galileo's search and rescue service also provides 'an important improvement of search and rescue systems [through] a combination of precise real-time positioning, reception of distress messages anytime nearly anywhere on earth, and … confirmation of reception to the user in distress' (Strodel *et al.* 2004: 8).

There are a number of maritime security challenges facing the EU (Commission 2006: 29–31), including illegal immigration by sea, smuggling and drug trafficking, terrorism, piracy and armed robbery at sea. In the case of smuggling and drug trafficking, for example, Griffiths and Jenks (2013: 37) suggest that containerised sea transport offers a simple, convenient and cost-effective mode of transport for drug smugglers. The UN Office on Drugs and Crime (UNODC 2008: 11) identify that many of the drugs entering Europe are transported by sea from the Caribbean and South America.

In addition, Asbeck (2009: 22–23) indicates that satellite images provided by the EU Satellite Centre, established in 2002 to support the Common Foreign and Security Policy, have been used to support European Union Naval Force (EU NAVFOR)–Atalanta operations around the coast of Somalia by identifying possible locations where pirates are operating.

Ships carrying liquefied natural gas, a highly flammable substance in its gaseous form, may be a potential terrorist target while at sea or in port (Testa 2004; McNicholas 2008: 248). Using tracking information, it is possible to identify whether a ship has followed the most appropriate route to its specified destination port, whether it has unexpectedly stopped at sea for any length of time without good reason (for example, due to mechanical failure), or if it has visited a different port to its notified destination without good reason. If there is no clear reason, then the ship may be inspected at sea or in port under the aegis of the International Ship and Port Facility Security Code (ISPS Code), a set of measures put in place to enhance ship and port security in the wake of the 9/11 attacks in the USA.[5] As a result of that event, the ISPS Code was implemented through Chapter XI-2 Special Measures to enhance maritime security under the 1974 International Convention for the Safety of Life at Sea (SOLAS Convention).[6]

As a result of the establishment of the European Border Surveillance System (EUROSUR) (Commission 2011c), information from satellites and from the SafeSeaNet system should contribute to both maritime security and maritime safety and enable the EU and national agencies to obtain surveillance information on EU external borders. However, in the case of illegal immigration, a particular problem in the Mediterranean region (UNODC 2011: 21), current systems are unlikely to help as the migrants often travel in small, overloaded boats that sink with consequent loss of life (Pugh 2004: 56). These small boats do not carry the Automatic Identification System on board and so the victims are unlikely to be rescued unless their boat sinks within sight of another vessel. This is supported by the Centre for Strategy and Evaluation Services, a multi-disciplinary research and consultancy service based in the UK, which identified a number of shortcomings relating to the use of satellite surveillance for maritime security, including a lack of wide-area maritime surveillance, only partial coverage of the open seas and the fact that surveillance systems have mainly been developed for maritime safety purposes (Centre for Strategy and Evaluation Services 2011: 5). This is an issue also identified by Coppini *et al.* (2011: 141) in relation to the use of satellite data collected for environmental monitoring.

Satellite systems and marine environmental protection

There are many examples of earth observing missions undertaken by ESA using satellites, including ENVISAT, the Earth Remote Sensing (ERS) satellites, the Meteosat European weather satellites and the MetOp polar orbiting satellites. These already play a significant role in the day-to-day functioning of the EU – for example, in satellite observation and satellite telecommunication links for vessel trafficking or as navigational aids.[7] As noted previously, radar equipment on board the ENVISAT satellite is used to support the routing of ships operating in ice-affected areas (ESA 2008).

Monitoring marine habitats to ensure adequate water quality and to maintain biodiversity is economically significant to EU prosperity with fisheries, tourism and other activities within 50 km of EU coastal states generating around €3.5 trillion (35 per cent) of the total EU GDP (Commission DG-MAF 2009: 3). Satellite data make a significant contribution to assessing the health of the marine environment through the monitoring of pollutants. They can also monitor naturally occurring biological hazards, as in the example of the ENVISAT images used to monitor algal blooms of cyanobacteria. Algal blooms can have a detrimental effect on aquatic plants, invertebrates and fish habitats, which poses a particular threat in the Baltic Sea (Paerl and Huisman 2008: 57).

Clark (1993: 361–365) identifies four categories of pollutants: physical, including sediments; chemical, including toxic, acidic and oily wastes; biological, including sewage and dissolved organic compounds; and thermal, where industrial discharge or the cooling requirements of power stations can raise the water temperature; oil pollution is well suited for investigation by remote sensing (Clark 1993: 362). It is one of the main pollutants for which satellite imagery has been used to protect the marine environment and there is a range of space and airborne remote sensing devices, with satellite sensors being used for preliminary oil spill assessment and airborne sensors for the more detailed analysis of oil spills (Jha *et al.* 2008: 236). Coppini *et al.* (2011: 152–153) concluded that satellite observing products provide robust operational information during oil spill incidents, more accurate predictions of oil spill drift and the identification of where oil may come ashore. Systems using optical sensors and synthetic aperture radar on board satellites can also aid in developing international action plans to clean up such spills.

As noted previously, CleanSeaNet is the EMSA service from which EU Member States can obtain (additional) satellite images to identify potential polluters and follow-up incidents using aerial surveillance. This covers all EU waters and images of possible intentional and accidental spills are produced several times a day.

CleanSeaNet data has been used in both the North Sea and Baltic Sea regions to support oil spill detection and monitoring activities since 2007. In the case of the North Sea, oil slick data had been collected in that region since the 1980s under the aegis of the Bonn Agreement[8] and, for the period 1986–2004, there was evidence of a decrease in the number of oil spills in the North Sea using aerial surveillance only (Carpenter 2007: 149–163). Data from aerial surveillance had also been collected since the 1980s in the Baltic Sea region under the aegis of the Helsinki Convention. In both regions, as no satellite observation data was available prior to 2004, monitoring was conducted using a system of aerial surveillance and was mainly confined to daytime hours.[9] Since 2004, both the Bonn Agreement and the Helsinki Commission annual reports[10] have included information on satellite observations, emphasising the increased accuracy of oil spill data and the ability to observe oil spills during the hours of darkness or in bad weather. Konstianoy *et al.* (2008: 71), citing the Finnish Environment Agency in 2004, indicates that there are around 10,000 oil spills in the Baltic Sea each year. The suggestion of the Helsinki Commission (2010: 36) that the Automatic Identification System should be used to backtrack from a detected oil spill and match a spill to a particular ship's track may potentially increase the likelihood that a polluter will be identified and prosecuted.

This illustrates the importance of satellite observation in that region and the benefit that can be gained from CleanSeaNet. In addition, CleanSeaNet operates within the International Charter for Space and Maritime Disasters[11] and, in the event of a major oil spill in EU and adjacent waters, this Charter provides a unified system for space data acquisition and delivery for those affected by both natural and human disasters (EMSA 2010).

The use of satellite imagery in conjunction with the CleanSeaNet service has already made a positive impact through increased levels of observation of spills in Europe's maritime areas. It has provided EMSA and Member States with more complete, accurate and verified oil spill information, allows observations to be made during the hours of darkness or periods of bad weather and increases the potential to identify and prosecute the owners of vessels that intentionally discharge oil at sea. This is significant as, based on data from EMSA (2011) that identified oil spills using satellite detection under CleanSeaNet for all EU Member States with a maritime border, the implication is that the vast majority of confirmed oil spills in recent year have been the result of intentional activities. The use of satellite images and synthetic aperture radar, when combined with the CleanSeaNet and SafeSeaNet services of EMSA, offers a valuable tool in identifying the sources of such intentional pollution and should also provide a deterrent against that activity as awareness of these systems increases.

Conclusions

Satellite technology and earth observation plays a significant part in supporting the operational tasks of EMSA. In the area of marine environmental protection, satellite observation provides a crucial tool for monitoring pollution. If we fail to monitor the marine environment to identify intentional pollution:

> [there is] the potential for intentional discharges of oil and other wastes into the [the waters of the North Sea] ... where a conscious decision has been taken to discharge wastes illegally ... from a perception that there is little risk of being identified as the source of such discharges.
>
> Carpenter (2007: 150)

Satellite observations of ocean currents, weather patterns and ice formation also play a role in ship safety, ensuring that vessels are not sailing into hazardous conditions and assisting with route planning.

Continued cooperation between ESA, EMSA and other EU agencies in policy areas relevant to the marine environment is vital to the continued security, safety and economic prosperity of the EU. Satellite technology can support a range of EU policy areas, from border security to identifying vessels that potentially pose a security threat, by monitoring areas where drug smugglers operate, or by identifying small boats potentially carrying illegal immigrants from West and North Africa. Satellite technology can also help to monitor the biodiversity of Europe's seas and ocean regions and to more widely monitor the impacts of climate change, providing the EU with evidence to support future actions and policies in those areas. It has been identified that:

> EMSA, the European maritime safety agency, has operational tasks in the field of oil pollution response, vessel monitoring, and the long-range identification and tracking of vessels. Space technology aids EMSA considerably in the day-to-day execution of these tasks, and particularly in the provision of maritime monitoring services to key stakeholders: the commission, EU member states and other EU bodies.
>
> Lourenco (2014)

This chapter, and the statement by Lourenco, highlight that there is clear evidence that the synergies offered between space, marine and other EU policy areas will greatly benefit cooperation between European Member States both at the current time and in the future.

Notes

1 For an overview of the sinking of the MV *Erika* and the subsequent actions of the EU, see Carpenter (2012: 260–263).
2 European Maritime Safety Agency – SafeSeaNet in Action. Available from: www. emsa.europa.eu/safeseanet-in-action.html. Last accessed December 2013
3 International Convention on the Safety of Life at Sea (SOLAS) (1974) as amended in May 2006. Available from: www.imo.org/About/Conventions/ListOfConventions/Pages/ International-Convention-for-the-Safety-of-Life-at-Sea-%28SOLAS%29%2c-1974.aspx. Last accessed 9 June 2015. The Long Range Identification and Tracking of Ships system is a requirement of the 1986 amendments of SOLAS at Chapter V. Available from: www. imo.org/OurWork/Safety/Navigation/Pages/AIS.aspx. Last accessed 9 June 2015.
4 For details of all the operational tasks of the EMSA, see www.emsa.europa.eu and select 'operational tasks'. Last accessed 9 June 2015
5 For further information on the ISPS Code, see www.imo.org/blast/mainframe. asp?topic_id=897. Last accessed 9 June 2015
6 For further information on the SOLAS Convention, see www.imo.org/About/Conventions/ ListOfConventions/Pages/International-Convention-for-the-Safety-of-Life-at-Sea-(SOLAS),-1974.aspx. Last accessed 9 June 2015. See also: www.imo.org/blast/mainframe.asp?topic_id=897. Last accessed 26 June 2014.
7 For details of ESA's Earth observing missions, see www.esa.int/Our_Activities/ Observing_the_Earth. Last accessed 9 June 2015.
8 Agreement for Cooperation in Dealing with Pollution of the North Sea by Oil and Other Harmful Substances 1983. Available from: www.bonnagreement.org/eng/html/ welcome.html. Last accessed 9 June 2015.
9 Convention on the Protection of the Marine Environment of the Baltic Sea Area, 1974. Available from: www.helcom.fi/helcom/en_GB/aboutus/. Last accessed 9 June 2015.
10 Available from: www.bonnagreement.org/eng/htm/welcome.html. Last accessed 9 June 2015. See also www.helcom.fi/action-areas/shipping/publications/ Last accessed 9 June 2015.
11 For further details of the International Charter for Space and Maritime Disasters Charter and charter activations, see www.disasterscharter.org/web/charter/home. Last accessed 18 December 2013.

References

Asbeck, F. (2009) 'EU Satellite Centre – a bird's eye view in support of ESDP operations'. *ESDP Newsletter* 8: 22–23.

Baumgartner, F.R. and Mahoney, C. (2008) 'The two faces of framing: individual-level framing and collective issue definition in the European Union'. *European Union Politics*, 9(3): 435–449.

Carpenter, A. (2007) 'The Bonn Agreement aerial surveillance programme: trends in North Sea oil pollution 1986–2004'. *Marine Pollution Bulletin*, 54(2): 149–163.

Carpenter, A. (2012) 'The EU and marine environmental policy: a leader in protecting the marine environment'. *Journal of Contemporary European Research*, 8(2): 248–267.

Centre for Strategy and Evaluation Services (2011) *Ex-post Evaluation of PAST Activities in the Field of Security and Interim Evaluation of FP7 Security Research: Maritime Security and Surveillance – Case Study*. Available from: www.cses.org.uk. Last accessed December 2013.

Clark, C.D. (1993) 'Satellite remote sensing for marine pollution investigations'. *Marine Pollution Bulletin*, 27(7): 357–368.

Commission (2002) Regulation (EC) No. 1406/2002 of the European Parliament and of the Council of 27 June 2002 Establishing a European Maritime Safety Agency. *Official Journal*, OJ L208/1–9, 5 August.

Commission (2003a) *White Paper – Space: a New European Frontier for an Expanding Union – An Action Plan for Implementing the European Space Policy*. COM(2003) 673 final, 11 November 2003.

Commission (2003b) *The European Union and the United Nations: The Choice of Multi-lateralism. Communication from the Commission to the Council and the European Parliament*. COM(2003) 526 final, 10 September 2003.

Commission (2005) *European Space Policy – Preliminary Elements. Communication from the Commission to the Council and the European Parliament.* COM(2005) 208 final, 23 May 2005.

Commission (2006) *Towards a Future Maritime Policy for the European Union. Commission of the European Union Green Paper*. COM(2006) 275 final, Volume II, ANNEX, 7 June 2006.

Commission (2007) *An Integrated Maritime Policy for the European Union. Communication from the Commission to the European Parliament, the Council, the European Economic and Social Committee and the Committee of the Regions*. COM(2007) 575 final, 10 October 2007.

Commission (2011a) *Bringing Space Down to Earth*. Available from: www.ec.europa.eu/enterprise/policies/space/index_en.htm. Last accessed December 2013.

Commission (2011b) *Towards a Space Strategy for the European Union that Benefits its Citizens. Communication from the Commission to the Council, the European Parliament, the European Economic and Social Committee and the Committee of the Regions*. COM(2011) 152 final, 4 April 2011.

Commission (2011c) *Proposal for a Regulation of the European Parliament and of the Council Establishing the European Border Surveillance System. Communication from the Commission*. COM(2011) 873 final, 12 December 2011.

Commission DG-MAF (2009) *The Economics of Climate Change Adaptation in EU Coastal Areas: Summary Report*. Luxembourg: Office for Official Publications of the European Communities.

Commission DG-MAF (2010) *Integrating Maritime Surveillance – Communication from the Commission to the Council and the European Parliament on a Draft Roadmap Towards Establishing the Common Information Sharing Environment for the Surveillance of the EU Maritime Domain*. COM(2010) 584 final. Luxembourg: Office for Official Publications of the European Communities.

Contzen, J.-P. and Ghazi, A. (1994) 'The role of the European Union in global change research'. *Ambio*, 23(1): 101–103.

Coppini, G., De Dominicis, M., Zodiatis, G., Lardner, R., Pinardi, N., Santoleri, R., Colella, S., Bignami, F., Hayes, D.R., Soloviev, D., Georgiou, G. and Kallos, G. (2011) 'Hindcast of oil-spill pollution during the Lebanon crisis in the Eastern Mediterranean, July–August 2006'. *Marine Pollution Bulletin*, 62(1):140–153.

Dudley, G. and Richardson, J. (1999) 'Competing advocacy coalitions and the process of "frame reflection": a longitudinal analysis of EU steel policy'. *Journal of European Public Policy*, 6(2): 225–248.

EMSA (2010) *Using Satellites to Improve Maritime Safety and Combat Marine Pollution: Prepared for Galileo Application Days 3–5 March 2010, Brussels. EMSA Galileo Brochure*. Available from: www.emsa.europa.eu/operations/cleanseanet/items/id/208. html?cid=122. Last accessed 9 June 2015.

EMSA (2011) *EMSA CleanSeaNet First Generation Report*. Available from: www.emsa. europa.eu/operations/cleanseanet/items/id/1309.html?cid=206. Last accessed 9 June 2015.

EMSA (2012a) *EMSA Annual Report 2011*. Available from: www.emsa.europa.eu/ publications/corporate-publications/item/143-annual-reports.html. Last accessed 9 June 2015.

EMSA (2012b) *SafeSeaNet in Action*. Available from: www.emsa.europa.eu/safeseanet-in-action.html. Last accessed December 2013.

ESA (2003) *Convention for the Establishment of a European Space Agency and ESA Council. ESA Convention, Pocket Version (English)*, 5th edn. Document Reference ESA SP-1271(E). Paris: ESA Publications Division.

ESA (2004a) *Benefiting our Economy: Navigation Through Sea Ice*. Available from: www.esa.int/Our_Activities/Observing_the_Earth/Benefiting_Our_Economy/Navigation_ through sea ice/(print). Last accessed 9 June 2015.

ESA (2004b) *Benefiting Our Economy: Optimising Shipping Routes*. Available from: www.esa.int/Our_Activities/Observing_the_Earth/Benefiting_Our_Economy/Optimising_ shipping_routes/(print). Last accessed 9 June 2015.

ESA (2007a) *ESA BR-269 Resolution on the European Space Policy: ESA Director General's Proposal for the European Space Policy, 30 June 2007*. Available from: www.esam-ultimedia.esa.int/multimedia/publications/BR-269/offline/download.pdf. Last accessed 9 June 2015.

ESA (2007b) *EU Unveils New EU Maritime Policy*. Available from: www.esa.int/Our_ Activities/Observing_the_Earth/EC_unveils_new_EU_maritime_policy/(print). Last accessed 9 June 2015.

ESA (2007c) *Civil Protection Assistance*. Available from: www.esa.int/Our_Activities/ Observing_the_Earth/Securing_Our_Environment/Pollution_tracking/(print). Last accessed 9 June 2015.

ESA (2008) *EU Maritime Day – Why is Space Relevant for Maritime Issues?* Available from: www.esa.int/Our_Activities/Observing_the_Earth/GMES/Why_is_space_relevant_ for_maritime_issues/(print). Last accessed 9 June 2015.

Griffiths, H. and Jenks, M. (2013) *Maritime Transport and Destabilizing Commodity Flows*. SIPRI Policy Paper 32. Stockholm: Stockholm International Peace Research Institute.

Helsinki Commission (2010) *Baltic Sea Environment Proceedings No. 123: Maritime Activities in the Baltic Sea – an Integrated Thematic Assessment on Maritime Activities and Response to Pollution at Sea in the Baltic Sea Region*. Helsinki: Helsinki Commission.

Hörber, T. (2012) 'New horizons for Europe – a European studies perspective on European space policy'. *Space Policy*, 28(2): 77–80.

Jha, M.N., Levy, J. and Gao, Y. (2008) 'Advances in remote sensing for oil spill disaster management: state-of-the-art sensors technology for oil spill surveillance'. *Sensors*, 8(1): 236–255.

Kohler-Koch, B. (2000) 'Framing: the bottleneck of constructing legitimate institutions'. *Journal of European Public Policy*, 7(4): 513–531.

Kostianoy, A., Litovchenko, K., Lavrova, O., Mityagina, M., Bocharova, T., Lebedev, S., Stanichny, S., Soloviev, S., Sirota, A. and Pichuzhkina, O. (2006) 'Operational satellite monitoring of oil spill pollution in the southeastern Baltic Sea: 18 months experience'. *Environmental Research, Engineering and Management*, 4(38): 70–77.

Lourenco, P. (2014) 'Space technologies can "support" the EU's maritime sector'. *The Parliament Magazine*, 24 July. Available from: www.theparliamentmagazine.eu/articles/news/space-technologies-can-support-eus-maritime-sector. Last accessed 9 June 2015.

Manners, I. (2002) 'Normative power Europe: a contradiction in terms?' *Journal of Common Market Studies*, 40(2): 235–258.

McNicholas, M. (2008) *Maritime Security: an Introduction*. Homeland Security Series. Oxford: Butterworth-Heinemann/Elsevier.

Mörth, U. (2000) 'Competing frames in the European Commission – the case of the defence industry and equipment issue'. *Journal of European Public Policy*, 7(2):173–189.

Paerl, H.W. and Huisman, J. (2008) 'Blooms like it hot'. *Science*, 320: 57–58.

Peter, N. (2004) 'The use of remote sensing to support the application of multilateral environmental agreements'. *Space Policy*, 20(3): 189–195.

Pugh, M. (2004) 'Drowning not waving: boat people and humanitarianism at sea'. *Journal of Refugee Studies*, 17(1): 50–69.

Strodle, K., Naddeo, G., Samson, J., Dieleman, P., Ferraguto, M., von der Hardt, H.J. and Gottifredi, F. (2004) 'System verification approach, methods and tools for Galileo'. Paper presented at ION 2003, Portland, OR, USA, 9–12 September 2003. Available from: www.reports.nlr.nl8080/xmlui/bitstream/handle/1092/570/TP-2004-136.pdf?sequences=1. Last accessed December 2013.

Testa, K. (2004). 'Are natural gas ships "boat bombs" for terror?'. *Associated Press*, 16 February. Available from: www.msnbc.msn.com/id/4276348/ns/us_news-security/t/are-natural-gas-ships-boat-bombs-terror/. Last accessed 9 June 2015.

UNODC (2008) *Drug Trafficking as a Security Threat in West Africa*. New York: United Nations.

UNODC (2011) *The Role of Organized Crime in the Smuggling of Migrants from West Africa to the European Union*. New York: United Nations.

Vogler, J. and Stephan, H.R. (2007) 'The European Union in global environmental governance: leadership in the making?' *International Environmental Agreements*, 7(4): 389–413.

Conclusion: 'To boldly go where no one has gone before'

Towards a European Space Strategy

Thomas Hörber[1]

Following the rich theoretical and empirical analyses in the previous chapters, it is now time to draw conclusions. Each chapter has brought its own perspective to the framing of space-related issues. Space has been shown to be a cross-cutting policy in a wide range of diverse EU policy fields. In Part II (polity), the authors analysed the institutions, with Thomas Hörber examining the role of the European Space Agency (ESA) and others addressing the EU: Emmanuel Sigalas on the European Parliament, Harald Köpping Athanasopoulos on the Council, Lucia Marta and Paul Stephenson on the Commission and Antonella Forganni on the Court of Justice of the European Union. Such a comprehensive analysis of the roles of the EU and other international institutions in defining space policy has not been published before.

Part III (politics) revealed the interests, expertise and entrepreneurship of different actors in the field of space policy. Ulrich Adam analysed the role of industrialists, Iraklis Oikonomou looked at trade unions, while Paul Stephenson shed light on the Commission as an actor with an infrastructure agenda. This section has also highlighted the impact that space policies have had on EU external relations. Christina Giannopapa, Maarten Adriaensen and Daniel Sagath examined the external relations of Member States with the ESA, while Veronica La Regina took a classical international relations perspective.

Finally, Part IV (policy) has highlighted a number of important European policies for which space technology already plays an important role. Jakob Feyerer analysed commercial policy and research and development in Galileo satellite navigation, Rik Hansen and Jan Wouters addressed industrial policy, while Angela Carpenter looked at the security and sustainable development dimensions of space applications for earth monitoring and observation.

Relevance of framing theory

All the authors have used framing theory as an analytical tool in their chapters. Framing has provided a unifying framework, allowing each author to think similarly and ask the same type of questions. Framing has remained sufficiently loose to hold the substantially different chapters in this book together. This lens was one of many that could have been used, such as integration theory or institutionalism, but, ultimately, was the lens that the editors felt would most empower the authors and equip them with a 'way of looking'. Recognising that framing is en vogue in European political analysis, the two editors hope that this edited book will put space policy on the map in European studies.

Kohler-Koch (2000: 516) asserted that framing theory pushes the limits of our theoretical understanding of the EU a bit further because supranational frames, such as space policy, create a new way of understanding what is beyond the nation state. Frames establish new norms and rules that make the political reality beyond the nation state comprehensible. A good example of this is the recent sanctioning of the ESA budget for the construction of Ariane 6 until 2020. Two frames were used to justify this expenditure, the frame of *freedom for Europe* to access space independently and the frame of space technology creating *employment in Europe*. These are clearly frames that no longer apply solely to national policies and which are shaping the development of a European space policy. The editors have not sought to push the boundaries of framing theory or to develop it further. Instead, the aim was to use it to provide a 'red thread' and a sense of cohesion to the book as a whole. Framing has proved to be an eminently useful methodology for studying this emerging field of space policy, which is so highly relevant to European integration. In doing so, this group of authors has made a case for space policy as a core contemporary feature of the European integration process.

Framing theory has also helped to structure the individual aspects of space policy, as developed in this book. However, the questions to be asked at this point are: 'Why it is only happening now?' and 'Why is space policy finally being recognised as a *European* policy?' One answer to these questions can be found in framing theory. The long-term historical policy studies that Dudley and Richardson (1999: 246) see as the only way to understand the development of frames are only now appearing in recent publications on space policy. Thus space policy is arguably not yet sufficiently understood by scholars as a driver of European integration in its own right, largely because space has not been studied as a 'policy domain', as have defence, environment, transport or even climate change.

Another reason for our lack of understanding of space policy as a European policy field may be found in the technical roots of space policy. It is true that the ESA has been around for decades, but experts in this field have always seen it as an essential part of European integration. Roy Gibson, the founding General Director of the ESA said, on the occasion of the celebration of the 50th anniversary of ESA on 7 May 2014 in Paris, that with the creation of ESA 'We were helping to construct Europe'.

Highly publicised events, such as the recent Rosetta landing on comet 67P/Churyumov-Gerasimenko on 6 August 2014, have finally caught the attention of the wider public. Such events have triggered wide media attention and secured great support and enthusiasm. The long-lasting effect, however, will most likely be limited again to technical experts, because once the event is over, media coverage falls away. That said, our fascination with the exploration of Mars and the mapping of its terrain will surely continue. However, without a strong and convincing frame (or frame set) for space – that is, an institutional setting and a 'narrative' that makes a case for space in Europe – sustaining the interest of the general public will be a challenge.

This book has tried to 'locate' space policy in European studies, with the hope that it becomes 'mainstreamed', attracting the interests of scholars who are busy researching in a wide range of closely related areas such as foreign policy, international relations, defence, security, environment, transport, migration, climate change, internal markets, trade and energy. Put more simply, space policy should no longer stand alone, but adopt a more definite (and definitive) position in the European integration process. One important analytical finding in that direction has been the use of the EU budget for major space endeavours such as Galileo. In the past, space researchers were working away in their corner, with little attention paid to them by political actors. However, with billions of euros now committed, politics come into play. Such large amounts of money need oversight and sanctioning by democratically legitimate institutions. The findings of this book reveal that Galileo was the game-changer, transforming the ESA's space expertise into a EU space policy, as shown in the development of policy frames on Galileo by the Commission (Chapters 6 and 13).

Jakob Feyerer (Chapter 13) analysed in detail the process of the Galileo project. Galileo was first framed as a public–private partnership with the logic that involving the private sector would help to save public money. After severe setbacks, the public–private partnership collapsed in 2007 and the Commission's framing of the project had to change, mainly to highlight European independence from the US global positioning system, but also to advocate Galileo's innovative character with the prediction of many positive spin-offs, such as cutting-edge technology and increases in productivity and employment.

Lucia Marta and Paul Stephenson (Chapter 6) looked at the inner workings of the Commission, including their framing of the Galileo venture. This chapter traced the way in which the European Commission has framed and reframed the issue of EU satellite navigation in recent years. It investigated how the agenda-setter has 'talked about' space policy, with a particular focus on Galileo, examining how the Commission's own institutional discourse has evolved over a decade through the use of 'frame sets'. In so doing, it was shown how the Commission chose to present the issue as politically and economically desirable for the EU. The earlier methodology chapter by Stephenson (Chapter 2) also looked closely at the language and vocabulary of EU legislation on Galileo.

Emmanuel Sigalas (Chapter 4) identified the engagement of the European Parliament in European space policy and the development of frames in that process. This chapter investigated how the Europeanisation of space has been portrayed as a legitimate development within European polity. It showed convincingly that the European Parliament has always been supportive of space policies. When it comes to framing this general policy choice, space has been portrayed as a highly promising policy field where Europe should reap multiple benefits. European independence has been shown to be the principal objective. It was argued that the European Parliament has acted strategically by adapting its space justification frames over time to ensure that its argumentation remains in line with the EU's various strategic priorities.

Harald Köpping Athanasopoulos (Chapter 5) looked at the Council of Ministers and its role in setting the EU's frames towards a European space policy. Rather than emphasising the Council's dominance as a legislator, the chapter argued that the Council has instead acted as the primary frame selector in the process of justifying the EU's ambitions to engage in space. In this context, the chapter concluded that human activities in space are increasingly seen as a means to some material end, rather than as an end in themselves. Particular attention was given in this chapter to an idealistic frame of European space policy, namely that space exploration ought to be prioritised because it provides us with immeasurable insights into the place and purpose of humanity in the universe.

For the first time ever, Antonella Forganni (Chapter 7) looked at the role of the Court of Justice of the European Union in European space policy. Unsurprisingly, there was no concrete case to be found because space policy only became an EU competence with the Lisbon Treaty of 2009. However, she did identify a number of Court of Justice of the European Union rulings in the field of aeronautics and telecommunications that are likely to develop into reference cases for future decisions on European space policies. In previous decades, the Court of Justice of the European Union has acted as a federator for the EU and for the internal market in particular, pushing frames pertaining to the free movement of services and goods. These are areas that may also be applied to space policies in the EU. It will be interesting to see whether the Court of Justice of the European Union develops a similar pro-integration dynamic, with space policy becoming more important for the functioning of the internal market.

The aforementioned chapters on the main European institutions make up the analysis of the EU polity in space policy. However, the EU is relatively new to space policy and the ESA predates the EU's engagement in this area. Therefore, two chapters were dedicated to the ESA: the first by Thomas Hörber on the development of the ESA from its predecessor organisation against the frame of independence for Europe, as well as technological innovation, while accepting the budgetary constraints that have been much tighter in Europe than with the other space powers (Chapter 1). The second chapter on the ESA focused on the development of its sound relationship with the EU (Chapter 3). Thomas Hörber outlined the discourse leading to a much closer relation between the ESA and the EU from the 2000s onwards. The main reason put forward was the growing political importance of space in European politics. This is shown in terms of growing budgets, as mentioned for Galileo, or politically, with the establishment of the ESA–EU Space Councils as joint sessions of their respective Ministerial Councils. The main frame behind this tendency would be *democratic legitimacy*. This does not mean that power politics will no longer interfere in these matters, but space policies are nonetheless growing in importance in the EU, which is an indication that a coherent European space policy is developing.

The chapters by Hörber (Chapters 1 and 3) and by Köpping Athanasopoulos (Chapter 5) also identified a lack of enthusiasm for further integration between the ESA and EU institutions in recent years, indicated by a decline in the importance of the Space Councils, previously seen as the main institutional innovation. Köpping Athanasopoulos recognised a certain jealousy on the part of the Commission over the influence of the Space Councils. The increasingly short sessions of the Space Councils, for which the Commission sets the agenda, are shown to be a consequence of the power politics between European institutions. The long-pending election of the Director-General of the ESA hampered innovation in European space policy. The ESA Convention does not give a fixed mandate and does not prescribe elections for the Director-General. As this is the most important political position in European space policy, it has led to horse-trading and bargaining among the Member States. Here the weakness of inter-governmentalism in the ESA can clearly be seen with the maintenance of the status quo. The issue was finally resolved with the election of ESA Director-General Johann-Dietrich Wörner.

Several chapters in this book have developed the message – perhaps the main argument of the book – that more supranationalism in European space affairs appears to be necessary. One last reason for the relative standstill in European space policy in recent years lies with the uncertainties over collaboration among the European institutions. Space is a new policy field and the actors are still in the process of deciding and asserting their positions. Beyond the tensions of power politics and intergovernmental arrangements, the question remains as to how to establish working channels of communication and collaboration between civil servants, industry and academics, as discussed in Part III of this book.

Ulrich Adam (Chapter 8) focused on the interests of industry using the example of the agricultural sector. This is relevant to our common focus given the link to the Common Agricultural Policy, which is still the most important spending line of the EU budget, and the modernisation of the agricultural sector through space technology. Finding widely accepted technology solutions that can raise productivity levels and promote environmental protection in farming has proved particularly challenging. Farming by satellite can help to achieve both sustainability and productivity growth in agriculture. What is commonly termed precision agriculture has emerged as a strategically important approach that can help to meet the challenge of feeding a growing world population. EU space policy, and therein the rapid growth of satellite applications in farm machinery, has presented the agricultural machinery industry with a unique opportunity to transform adverse and outdated views about agriculture and agricultural machinery in EU policy-making. The agriculture industry is shown to use space technologies deliberately to reframe its image, outlook and contents in EU policy-making.

Iraklis Oikonomou (Chapter 10) looked at the interests of trade unions in European space policy with the example of the European Metalworkers' Federation. He considered past framing activities very critically – that is, Mörth (2000) in EU defence policy integration and Stephenson (2012) in EU space policy. For him, framing poses an inherent problem, namely, the lack of a theory of *power* embedded in it, or, in other words, the ontological primacy of discourse and ideas over material interests. This chapter provides an analysis of the discursive patterns articulated by the European Metalworkers' Federation in the interrelated policy fields of space and defence, arguing that framing cannot account for the discursive regularities and the maintenance of a stable, pro-industrial stance at the core of labour's rhetoric. Such a set of patterns is instead the outcome of the exercise of hegemony by the ruling social and economic forces of the internationalised aerospace industry. The chapter concluded that the defence and space industries exercise an ideological dominance to which trade unions themselves submit – for example, by safeguarding employment.

Paul Stephenson (Chapter 9) identified the great European infrastructure projects of the Trans-European networks (TENs) as the Commission's agenda to complete the Single Market in 1992 and to increase externally the competitiveness of the EU in the world. TENs were shown to be one of the main tools for managing progress of the EU, with space is becoming increasingly important in this context. Thus the frame of infrastructure development was extended to include space technologies and the consequent institutional connections to the ESA. TENs were justified using three different frame sets. Their rapid development, against the backdrop of 30 years of neglect in European transport policy, arguably helped to create a policy-making environment and secured an ideational consensus that would subsequently allow for Galileo to be implemented. Transport policy can thus be seen to be complementary to space policy, while also proving a European policy-making foundation.

In international politics, the growing assertion of the EU as an actor also features European space policy as a tool. What are internal and what are external politics is not so easy to define in the EU context. Therefore, Christina Giannopapa, Maarten Adriaensen and Daniel Sagarth (Chapter 11) took a look at external politics from the perspectives of the nation states. Such an approach has its merits because the ESA has 20 Member States with a variety of governance structures, strategic priorities and motivations regarding their space activities, which, in turn, shape the industrial structure and, arguably, vice versa. A number of countries engage in space activities exclusively though the ESA, while others have their own national space programme. Some consider the ESA as their prime space agency, while others have, in addition, their own national agency. The chapter provides an up-to-date overview of national space governance structures, strategic priorities and the industrial capabilities of the ESA Member States.

Veronica La Regina (Chapter 12) takes a true international relations perspective, arguing that space is at the heart of Europe's international relations with the rest of the world, involving the Member States of the ESA and the EU. As such, she argues that analysing the framing of space policy is complex because of multiple, often competing, internal and external political interests and values. This chapter sought to identify the dynamics of the framing process to better understand the weaknesses of EU institutional reforms. Different framing policy options are expressions of different interests, such as market, defence, advocacy coalitions and individual policy entrepreneurs. The chapter presented the different framing approaches behind the EU's international cooperation with third countries and/or overseas regional blocs. It identified two main frames: geopolitics and industrial policy. The first frame is typically adopted in a political context, dealing with countries active in space exploration, where cooperation ranges from mere coordination actions to the sharing and managing of space-based assets. The second frame operates when cooperating with third countries without space-based technologies and where the EU attempts to promote its own industrial interests to gain a greater share of overseas markets.

Rik Hansen and Jan Wouters (Chapter 14) examined the development of an EU industrial policy for the space sector. The rise of the EU as a space actor, especially as an initiator, then as owner and operator of large-scale programmes such as Galileo and Copernicus, has raised a number of questions with regard to industrial policy. Based on the experiences from the procurement of Galileo and on present discussions on space industrial policy within the EU, the chapter found that the EU's political ambitions in space are reasonably well-defined, but that the specific policy tools and legal instruments to put them into practice are far from complete. The chapter analysed the frames used in the discourse of the European Commission and the European Space Council on the topic and concluded that it is high time for a European industrial space policy.

Finally, Angela Carpenter's policy analysis of Galileo and Copernicus (Chapter 15) showed that satellite technology offers many benefits to Europe's maritime regions and its marine environment, ranging from tracking shipping to monitoring pollution and enforcing border controls. This chapter made the crucial link to prominent EU policies, such as sustainable development, that need space technologies to ensure their proper functioning, which explains, at least to some extent, the increasing importance of space technologies in the EU.

These chapters have all strengthened the case for an increasing importance of space aspects in the future, despite teething problems – for example, in Galileo. What remains is to put the findings into further context and to link them with the broader remit of the book as outlined in the introduction.

The case and scope for a European space policy

It was argued in the introduction that the early essential ideals of peace and prosperity have been largely fulfilled and that the development of a European space policy could well have the potential of giving the European integration process new impetus, which might, in turn, provide a driving force for the original ideals of peace and prosperity in Europe. This rationale is based on what may seem a rather naïve assumption that we do actually have the power to decide where to direct the European integration process. The chapters provide good evidence that the development of a European space policy has an outreach potential that could give new impetus to European integration. This book has shown that 'space policy' cannot de defined by a set of directives and regulations that emanate from a 'DG Transport and Space'. The addition of space to the name is nevertheless new, and telling of the growing importance of space in European politics. The transversal nature of space policy has the potential to drive the European integration process forward in the future. Given the incontestable achievement of economic advancement and the resulting economic strength of today's EU, it is not unreasonable to expect that European economies might support a space programme that becomes a common European policy and which could be more sophisticated than what we see today under ESA auspices. Such a European space policy would combine the two factors outlined earlier for all the major European integration initiatives – that is, vision and practical implementation. Closer integration between the Member States seems to be the logical outcome.

With peace and prosperity largely achieved, it may now be time to look for new objectives for the EU, externally and internally, in the twenty-first century. Internally, a consolidation process already seems to be under way. The question of defining the borders of Europe will be vital in that context (Berg Eriksen 1997: 111–117). Moreover, the project of a European Constitution – amended into the Lisbon Treaty – is another indication that Europe is also consolidating itself as a law- and rule-based polity. Thus *consolidation* seems to characterise the internal objectives of the EU in the twenty-first century. The increased importance of space policy in the institutional architecture of the EU is contributing to this consolidation process, as seen with the newly created DG Transport and Space and the ESA–EU Space Councils. As for the external objective, the elimination of armed conflict in Europe may be regarded as an *acquis*, except in its border regions, such as Ukraine, but certainly within the EU. Space exploration, with its concrete application in the European space policy could become this future external objective of the EU. That leads to the following question: 'Why space and why in a European Framework?' Space exploration has been shown to have significant outreach potential. Space exploration could reflect the creative dynamism of the European integration process. Monnet's argument still stands that the supranational model is best applied in areas where technological breakthrough opens up new fields of activity, provided that it does not touch on national sensitivities as, for example, in the fields of nuclear research or military security (Monnet 1976). A space programme at the cutting edge of technological development clearly fits this description. In the past, a European space policy has never featured very high on the agenda of any individual European Member State or, indeed, on the agenda of the EU itself, which takes out the one caveat of 'national sensitivities'. Some of the chapters in this book have made a case for the growing importance of space affairs in European politics and one might even argue that this is the raison d'être of this book. However, the ESA has arguably never been at the core of the European integration process. This may change in the future. The combination of 'low politics' and 'novelty' suggests that a European space policy is ideally positioned to become a key feature of the EU's future policy objectives.

Institutional considerations

Against this background, Monnet had a point when he said that no bold endeavour can succeed under unanimity rule (Monnet 1976: 333, 413). Some thought should therefore be given to the idea of applying supranational principles to the ESA. Considering the intergovernmental origins of the ESA, it is important to point out that different forms of majority voting already exist under the current ESA Convention for the ESA Ministerial Council meetings (ESA Convention, 2010: Article XI). However, the ESA remains an intergovernmental institution because its constituent members in the decision-making body are Member States of the Council of Ministers. As such, it only enjoys indirect political legitimacy through the governments of the Member States. The major supranational institutions belong to the EU; the European Parliament has direct democratic legitimacy and the Commission possesses a collegiate cabinet working for the interests of Europe, not the Member States, just like the Court of Justice of the European Union. It is the argument of this book that, with the growing importance of European space policy, increasing political legitimacy is needed – for example, the Galileo budget of several billion euros needs the budgetary supervision of an institution with direct political legitimacy – that is, the European Parliament and, arguably, the auditing of the European Court of Auditors – to examine questions of effectiveness and value for money for the European taxpayer. This question of political legitimacy is the most important reason for the debate about the relationship between the ESA and the EU and the question how the ESA should evolve in the future. More supranational governance and less intergovernmental procedure is what may provide this legitimacy to the ESA. On the basis of greater political legitimacy, the ESA's well-established expertise in space technology can be fully used. It would seem natural for it to become the executive body of a European space policy if the political institutions remain with the EU, not least because the EU has not proved capable of handling major technology projects such as Galileo well (see Chapter 13). At this point, the issue of the institutional independence of the ESA (see Chapter 3) comes up. Institutional arrangements and indeed funding for a European space policy will become important when the benefits of space exploration become more obvious. This institutional evolution of the ESA will be key if it is to manage and steer effectively the increased importance of future space policy. The consequences of increased activity and status are difficult to gauge, but, as Chapters 11 and 12 have shown, other space powers are bringing about institutional change to be able to make these important first steps into space. Closer cooperation between the ESA and the EU should also allow Europe to continue into that direction.

Outlook

Having argued for a common space policy in a European context, we might consider comparable historical experiences and how they apply to the challenges the EU faces in space exploration. Such comparisons help to provide an outlook of what lies ahead and what further research may provide insights for future European space policy. What are the remaining political and practical challenges (financial, scientific and organisational)? Where should the study of European space policy go?

Future research will need to consider whether adequate guiding principles for space exploration are in place or how they might be developed to guarantee the responsible use of the outcomes of European space policy – for example, preventing exploration turning into exploitation. European studies – as a discipline at the crossroads of social sciences, law and history – is ideally poised to provide a framework for such studies and, at the same time, to develop a theoretical basis for a European space policy with a view to the key existential questions of the European integration process: What does Europe stand for? What does it do for me? And where should it go?

'To boldly go where no one has gone before' is the title of this conclusion, with a nod to *Star Trek*, for all its popular idealism, imagination and vision. The European integration process will need such an idealistic drive to answer the fundamental questions around its raison d'être, as outlined in the introduction, conclusion and several chapters of this book. Clearly, supranational principles have provided the basis for such dynamism in the European Communities and, later, the EU. Against this background we might look to future space endeavours being led under the unifying, protective and incentivising common umbrella of the EU, with the ESA in charge of implementing a burgeoning European space policy. The findings of this book to some extent reflect the dynamic and supranational aspirations of the authors for a future European space policy. However, current political developments are not necessarily moving in that direction, as seen at the last ESA Ministerial Council on 2 December 2014 in Paris. Despite the preference for a close and constructive relationship with the EU, in the final Resolution the ESA Member States expressed:

> ...their clear preference for a relationship between ESA and the European Union which:
>
> (a) keeps ESA as an independent, world-class intergovernmental space organisation according to the Convention, pursuing its role in the framework of the implementation of programmes decided and funded by Member States and conducted on their behalf; and
> (b) makes ESA the long-term partner of choice for the EU for jointly defining and implementing the European Space Policy together with their respective Member States, in a dynamic and stepwise approach on the basis of the 2004 Framework Agreement, with regular meetings of the Space Council.
>
> (ESA 2014: 5)

The ESA seems to prefer the status quo while supporting further step-by-step integration. Incrementalism has been a feature of past European integration. Let us hope that developments will move fast enough to allow Europe to meet the challenges facing it by developing an ambitious, politically backed European space policy.

Note

1 For the key argument made in this conclusion, see Hörber (2006).

References

Berg Eriksen, T. (1997) 'The European self-image'. In: J.P. Burgess (ed.), *Cultural Politics and Political Culture in Postmodern Europe.* Postmodern Studies No. 24. Amsterdam: Rodopi.

Dudley, G. and Richardson, J. (1999) 'Competing advocacy coalitions and the process of "frame reflection": a longitudinal analysis of EU steel policy'. *Journal of European Public Policy*, 6(2): 225–248.

ESA (2014) *Resolution on Europe's Access to Space*. ESA(2014)/C-M/CCXLVII/Res. 3 (Final), adopted 2 December 2014.

Hörber, T.C. (2006) 'To boldly go where no one has gone before: the development of a European Space Strategy (ESS)'. *L'Europe en formation*, 2/2006: 19 28.

Kohler-Koch, B. (2000) 'Framing: the bottleneck of constructing legitimate institutions'. *Journal of European Public Policy*, 7(4): 513–531.

Monnet, J. (1976) *Mémoires*. Fayard, Paris.

Mörth, U. (2000) 'Competing frames in the European Commission – the case of the defence industry and equipment issue'. *Journal of European Public Policy*, 7(2): 173–189.

Stephenson, P.J. (2012) 'Talking space: the European Commission's changing frames in defining Galileo'. *Space Policy*, 28(2): 86–93.

Index

Page numbers in *italics* denote tables, those in **bold** denote figures.